Statistical Analysis of
Weather Modification Experiments

LECTURE NOTES IN STATISTICS

A Series Edited By

D. B. OWEN

Department of Statistics
Southern Methodist University
Dallas, Texas

Vol. 1 Incomplete Block Designs, *by Peter W. M. John*
Vol. 2 Matrix Derivatives, *by Gerald S. Rogers*
Vol. 3 Statistical Analysis of Weather Modification Experiments,
edited by Edward J. Wegman and Douglas J. DePriest

Additional Volumes in Preparation

Cloud seeded by Dr. Shelton Elliott with one-half gram of silver iodide. The photograph was taken 7 minutes after the left-hand side was seeded and shows the left-hand side snowing. The right-hand side remains unaffected although later circulation within the cloud caused precipitation from the entire cloud. The photograph was taken on March 8, 1966 over Panamint Valley, California and was provided by the Naval Weapons Center, China Lake, California.

Statistical Analysis of Weather Modification Experiments

edited by

Edward J. Wegman and Douglas J. DePriest

Statistics and Probability Program
Office of Naval Research
Arlington, Virginia

MARCEL DEKKER, INC.   New York and Basel

Library of Congress Cataloging in Publication Data

Workshop on the Design and Analysis of Weather Modification Experiments, Florida State University, 1978
   Statistical analysis of weather modification experiments.

   (Lecture notes in statistics ; v. 3)
   Papers presented at a workshop held Oct. 26–27, 1978 at Florida State University.
   Includes index.
   1. Weather control–United States–Experiments–Congresses. 2. Weather control–Statistical methods–Congresses. I. Wegman, Edward J. II. DePriest, Douglas J. III. Title.
IV. Series: Lectures notes in statistics (New York, N.Y., 1980–   ) ; v. 3.
QC928.7.W67   1978   551.68   80-24352
ISBN 0-8247-1177-7

COPYRIGHT © 1980 by MARCEL DEKKER, INC. ALL RIGHTS RESERVED.

Neither this book nor any part may be reproduced or transmitted in any form or by any means, electronic or mechanical, including photocopying, microfilming, and recording, or by any information storage and retrieval system, without permission in writing from the publisher.

MARCEL DEKKER, INC.
270 Madison Avenue, New York, New York 10016

Current printing (last digit):

10 9 8 7 6 5 4 3 2 1

PRINTED IN THE UNITED STATES OF AMERICA

Preface

Weather modification is a phrase that stirs up controversy at several levels. At the most philosophical level, there is a legitimate question as to whether or not human beings ought to be tampering with meso- and macro-scale environmental phenomena with strong implications on ecological balances in nature. Such issues have not yet been widely addressed because of the lack of answers to more fundamental technical questions: "Can the weather be effectively modified?" and if so, "Can these modifications be controlled?" The answers to these questions have not always been clear-cut and there appear to be excessive claims on all sides of the controversy. A positive answer to both questions would have a profound impact in many sectors of public life. Causing rain in drought-stricken regions, especially in agricultural states, can have enormous impact on the economy. Preventing severe thunderstorms, tornadoes, and hail can have an equally beneficial effect. On a global scale, the prevention and reversal of the desertification process by artificially enhanced rainfall has implications beyond mere economics. From another point of view, it is clear that naval and amphibious operations can succeed or fail depending on weather conditions, and control of these offers a tactical and strategic advantage to the controlling side.

With so much at stake, the answers to our two fundamental questions take on new import. They are not merely scientific curiosities, but questions with profound ecological, economic and ethical significance. Can the weather be modified? Most meteorologists seem to answer this question with a resounding "yes," but with a caveat, "under the right conditions." Unpredictability, however, seems to be the hallmark of weather and, hence, there is a natural role for statisticians. Statisticians are on the whole natural-born skeptics and their answer to this question, based on data provided by meteorologists, is much more equivocal. As editors of this volume, we throw our lot with those who say, "yes under the right conditions." We have included in the frontispiece a photograph of a cloud, the left half of which has been seeded. The left half is snowing, the right half remains unaffected.

Cloud seeding and weather modification are enormously complex affairs. Even granting that seeding causes physical changes in the clouds, many questions remain. Can it be made to rain in specific locations at specific times? What is the overall impact on the water balance, say over an annual cycle? Does rain enhancement here cause rain deficit elsewhere? Are there effects which are felt at long range from the seeding target sites? In short, can weather modifications be controlled? This, it seems to us, is the real question.

While both cloud seeding and statistical technology have been making steady improvements, the question of controllability is still equivocal. We have organized the present volume (and the workshop from which it arose) to address, at least partially, the question of controllability. The unifying theme of all the papers in this volume is a weather modification experiment called the Santa Barbara Convective Seeding Test Program, which was conducted by North American Weather Consultants (NAWC) from 1967 through 1971 for the Naval Weapons Center (NWC), China Lake, California. This experiment was conducted by Dr. Robert Elliott of NAWC under the sponsorship of Dr. Pierre Saint-Amand of NWC. It is fair to characterize them as the protagonists. Professor Jerzy Neyman has maintained a long-standing interest in weather modification, has written extensively on the topic and has been noted as a leading skeptic of some of the claims of the weather modifiers. Professor Bradley and his colleagues at the Florida State University came onto the scene as observers, but as experts in data analysis techniques who would, we hoped, provide an unbiased viewpoint! Professors Kempthorne, Court and Gabriel served in the role of discussants with various shadings of expertise and points of view. This cast of characters, representing widely differing views, was brought together with the expectation that lively discussion would result. We were not disappointed with the outcome. To use a phrase of one of our more diplomatic colleagues at the Office of Naval Research (ONR), the discussion was "colorful."

While the heat of the moment may be tempered by cooler reflections, the diversity of points of view is reflected in the written papers. Professor Neyman and Dr. Saint-Amand take somewhat philosophical points of view, with the remaining contributions more technical in nature. It is our hope that the juxtaposition of these differing points of view makes not only for interesting reading, but also for a clarification of the issues involved.

Many people deserve to be acknowledged for their contributions to this volume. The workshop entitled *A Workshop on the Design and Analysis of Weather Modification Experiments*, from which this volume arose, was formulated in the heat of the desert as we travelled from NWC to Los Angeles. We had just visited Dr. Saint-Amand at NWC and earlier, Professor Neyman at Berkeley. Their differing points of view were the immediate inspiration for the workshop. (The heat of the desert also made improved prospects for rainmaking seem very desirable.) However, for several years earlier the Statistics Program at ONR, under the leadership of Dr. Bruce McDonald, had sponsored

Preface

research to develop improved methodology for the statistical evaluation of weather modification experiments. This work was initiated and developed in close cooperation with the Atmospheric Sciences Program at ONR and the Environmental Surveillance Program at the Naval Air Systems Command. Part of this research was under the direction of Professor Ralph Bradley, and we are grateful to him for not only his excellent scientific contributions, but also for allowing himself to be pressed into service as host and local arrangements coordinator. The final program was arranged by Professor Bradley and Dr. DePriest. Finally, we are grateful to our secretary, June Sturdivant, who often bears the brunt of the workload of our projects.

Edward J. Wegman
Douglas J. DePriest

Contents

Preface   iii

Contributors   ix

Developments in Probability and Mathematical Statistics Generated by Studies in Meteorology and Weather Modification   1
    Jerzy Neyman

Some Operational Considerations in Evaluation of Weather Modification Programs: A Short Excursion into Epistemology   11
    Pierre Saint-Amand

Some Approaches to Statistical Analysis of a Weather Modification Experiment   33
    Ralph A. Bradley, Sushil S. Srivastava and Adolf Lanzdorf

Comparing the Testing of Hypotheses Based on Lognormal and Gamma Distributions   55
    M. Hanson and L. Barker

A Multivariate Methodology for the Analysis of Weather Modification Experiments   61
    Elton Scott

Physically Meaningful Covariates   79
    Robert D. Elliott

Some Statistical Aspects of Weather Modification Studies 89
  Oscar Kempthorne

Limitations of Statistics in Weather Modification 109
  Arnold Court

Comments on the Reanalysis of the Santa Barbara II Cloud Seeding 113
Experiments
  K. Ruben Gabriel

Comments on the Discussion at the Workshop on the Statistical 131
Design and Analysis of Weather Modification Experiments
  Jerzy Neyman

Response to the Discussion 139
  Ralph A. Bradley

Index 143

Contributors to the Workshop

Lawrence Barker, Department of Statistics, The Florida State University, Tallahassee, Florida

Ralph A. Bradley, Department of Statistics, The Florida State University, Tallahassee, Florida

Arnold Court, Department of Geography, California State University, Northridge, California

Robert Elliott, North American Weather Consultants, Goleta, California

K. Ruben Gabriel, Department of Statistics, University of Rochester, Rochester, New York

Morgan Hanson, Department of Statistics, The Florida State University, Tallahassee, Florida

Oscar Kempthorne, Statistical Laboratory, Iowa State University, Ames, Iowa

Adolf Lanzdorf,[†] Department of Statistics, The Florida State University, Tallahassee, Florida

Jerzy Neyman, Statistical Laboratory, Department of Statistics, University of California, Berkeley, California

---

[*]*Current address:* North American Weather Consultants, Salt Lake City, Utah
[†]*Current affiliation:* Strategic Planning Associates, Washingon, D.C.

Elton Scott, Department of Statistics, The Florida State University, Tallahassee, Florida[‡]

Sushil Srivastava, Department of Statistics, The Florida State University, Tallahassee, Florida

Pierre Saint-Amand, Earth and Planetary Sciences Division, Research Department, Naval Weapons Center, China Lake, California

---

[‡]Permanent address: Finance Department, School of Business, The Florida State University, Tallahassee, Florida

Developments in Probability and Mathematical Statistics Generated by Studies in Meteorology and Weather Modification

Jerzy Neyman

Statistical Laboratory
University of California
Berkeley, California

Abstract

Invariably, a persistent study of a category of natural phenomena generates novel mathematical developments, occasionally including sophisticated mathematical concepts. Of the mathematical developments generated by meteorology and efforts to modify weather, the most sophisticated seems to be the concept of characteristic functional. It was first published by A. N. Kolmogorov to be unnoticed for twelve years. It was reintroduced in 1947 by L. M. LeCam, inspired by studies of meteorology. Substantially less sophisticated mathematical concepts stem from cloud seeding experiments. They include (i) "outlier prone" and "outlier resistant" distributions, (ii) two mechanisms of response to cloud seeding, and (iii) the concept of variability of response to cloud seeding.

1 INTRODUCTION

As is well known, prolonged studies of any substantial domain of empirical science generate novel developments in mathematics. Frequently, these new mathematical developments are of only a modest level of sophistication. On occasion, however, they are quite impressive. Meteorology in general and rain stimulation in particular, are no exceptions to the above general rule. The purpose of the present article is to describe several cases that came to my attention.

2 CONCEPT OF CHARACTERISTIC FUNCTIONAL

To my knowledge, chronologically the first, and also the first in the degree of mathematical sophistication, is the work of Lucien M. LeCam. The concept of characteristic functional was inspired by studies of meteorology as may be seen from the quotation below. We include the title and first few lines of LeCam's (1947) in the *Compte Rendus* of the Paris Academy.

> **CALCUL DES PROBABILITES.** — *Un instrument d'étude des fonctions aletoires: la fonctionnelle caractéristique.* Note de M. Le Cam présentée par M. Emile Borel.
>
> En vue de préciser et généraliser certains résultats obtenus par M. Halphen dans l'étude statistique des débits, nous avons été amené á introduire un instrument de calcul qui presente pour les fonctions aléatoires le méme intérét que la fonction caractéristique pour les variablesé aléatoires.

At the time, LeCam was a student at the University of Paris and earned his living by working at the Electricité de France. His main contact was E. Halphen, who was interested in the relationship between rainfall and river runoff, a domain of meteorology that continues to be important to this day.

While the quotation above and the entire initial paragraph of LeCam's Note document the fact that the idea of characteristic functional came to LeCam's mind from studies in meteorology, the priority in the development of the concept belongs to A. N. Kolmogorov. I am grateful to Professor Grace Yang for calling attention to a passage in the book of Theodore E. Harris (1963), p. 54, which is reproduced below. This passage illustrates the somewhat explosive effects of LeCam's Note, even though the concept of the characteristic functional was published by Kolmogorov twelve years earlier. The conceptual delicacy of the idea must have been well above the general level of contemporary probabilists. In comparison, other meteorology-generated mathematical-statistical developments described below are somewhat pedestrian.

5. Moment-generating functionals

The usefulness of the moment-generating function in treating scalar and vector random variables suggests finding an analogous device for dealing with random point-distributions. We shall define a *moment-generating functional*, which we shall sometimes abbreviate by MGF (as opposed to mgf for moment-generating function).

The MGF is a modification of the *characteristic functional* introduced by LeCam (1947) for random point functions and by Bochner (1947, 1955) for random set functions. Bartlett and Kendall (1951) demonstrated the importance of such functionals in applications; see also Bartlett (1955), where a number of applications are discussed. Bochner (1955) applies the term *generating functional* to what we shall call the MGF. Bochner gives a general treatment of characteristic functionals but the treatment to be given here seems best suited for our needs.[1]

---

[1] Added in proof: According to Prohorov (1961), the characteristic functional was employed by Kolmogorov (1935) for distributions in a Banach space.

3 OUTLIER PRONE AND OUTLIER RESISTANT PHENOMENA AND DISTRIBUTIONS

One of the classical fields of applications of statistics is the evaluation of measurements, unavoidably subject to errors. Here, the basic presumption is that the error in a measurement, say of a coordinate of a star, is a sum total of a number of independent factors, such as motions of the atmosphere in front of the telescope, the vagaries of temperature within the observatory, inaccuracies of readings, etc., etc. The familiar central limit theorem on probability suggests then that the total error in a measured coordinate must be approximately normally distributed. Here, then, the probability of an error exceeding a substantial multiple of its standard error is negligibly small. If such an error does occur, the presumption is that it must be due to some extraneous cause, such as a mistake in copying, some "gross error," an "outlier," to be ignored.

This kind of consideration created a substantial literature, directed towards an efficient methodology for "cleaning" gross errors from observational data. This literature is exemplified by papers by Ferguson (1961) and by Grubbs (1969).

Long experience in studying rain stimulation experiments indicates that the distribution of nonzero rainfall per experimental unit (an experimental day or a storm) is J-shaped, with a long "tail" and with frequent gaps towards the end of the "inverted J." Figure 1, reproduced from Neyman and Scott (1971), illustrates what we not call the "outlier prone" character of rainfall. While it refers to a particular locality, Little Egypt (Illinois), this figure is typical. One of the histograms summarizes the observations on "air-mass" and the other on "non-air-mass" storms. It is seen that the air-mass storms are more outlier prone than the opposite stratum. The same paper of Neyman and Scott (1971) contains quite a number of illustrations similar to Figure 1, referring to a variety of conditions.

These empirical findings lead to two important conclusions. One is that the rejection of rainfall data relating to a particular day merely because they appear as "outliers," exposes the evaluation of the effects of seeding to the danger of being unrealistic and, therefore, misleading. The other conclusion is that the mathematical aspect of the situation is in need of two new concepts: "outlier-prone" and "outlier-resistant" families of distributions. These two concepts have been introduced in Neyman and Scott (1971). Also, see papers by Richard Green (1974 and 1976).

Out of the broadly familiar families of distributions the normal and the Cauchy are outlier resistant. The outlier-prone families may be exemplified by the log-normal and the gamma distributions. However, the log-normal distribution is not J-shaped and, therefore, does not appear promising as a good fit for precipitation data. One the other hand, our long experience indicates that the nonzero rainfall per experimental unit of fixed duration (like 24 hours, etc.), falling in a fixed "target" (that is, rainfall measured by a network of gages of fixed locations) is satisfactorily fitted by the gamma distribution with density

$$p_X(x) = \frac{\delta^\gamma}{\Gamma(\gamma)} x^{\gamma-1} e^{-\delta x} \qquad (1)$$

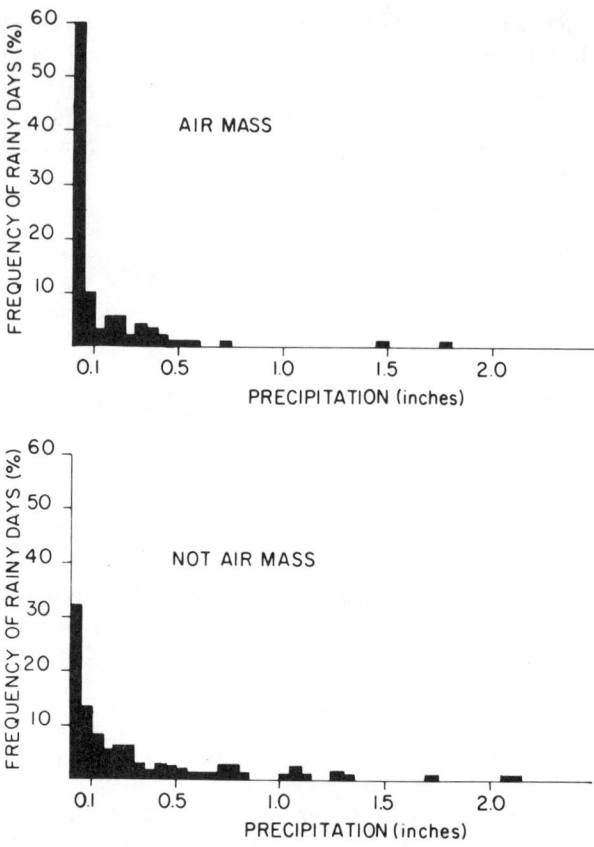

Figure 1 "Outlier prone" distributions of nonzero precipitation amounts. Reprinted from Neyman and Scott (1971) with permission of Academic Press.

Here $\gamma > 0$ is the shape parameter and $\delta > 0$ the reciprocal of the scale parameter. Furthermore, the same long experience indicates that, while the seeding of clouds can change the value of the scale parameter, it does not seem to affect the value of the shape parameter. In consequence, our customary methodology of evaluating cloud seeding experiments, with experimental units of fixed duration and with fixed target, is based on formula (1), with only the parameter δ expected to be affected by seeding.

4 TWO MECHANISMS OF RESPONSE TO CLOUD SEEDING

The distributions of rainfall per experimental unit (say, per one experimental day) are somewhat peculiar: the probability, say $1 - \theta$, of the rainfall being exactly zero is positive. Cases are on record where the seeding of clouds over a conventionally sized target appears to "trigger" the rainfall that would not otherwise fall. Here, then, we visualize that the probability θ has been increased. However, some other experimental results suggest that the seeding decreased θ.

In an effort to treat these observations mathematically, it is necessary to introduce the concept of a special mechanism, say M_1, that governs the change in the value of θ due to seeding.

The same kind of observations indicate the necessity of introducing another mechanism, say M_2, that governs the effects of seeding "per wet day." Here again the mechanism M_2 may tend to increase or to decrease the target precipitation.

The above circumstances are discussed in several papers published in Volume V of the *Proceedings of the Fifth Berkeley Symposium on Math—Stat. and Prob.* (University of California Press, 1967, Berkeley, CA 94720). One of them (Neyman and Scott, 1967b) is concerned with the design and evaluation of a single unrestrictedly randomized experiment. The visualized general situation is as follows.

The experimenting meteorologist has an idea about the weather conditions "suitable" for cloud seeding (e.g., if there is practically no moisture in the atmosphere, it would be useless to attempt to bring rain!) Let $\{U_n\}$ designate the sequence of experimental days considered "suitable." When such a day arrives, a randomized decision has to be made whether to proceed with seeding or not. This is done by tossing a figurative coin with a preselected probability, say π, of falling heads. (Actually random number generators are used for this purpose.) If the coin does fall heads, there is seeding. Otherwise, the corresponding experimental unit is assigned to be a control unit.

Let the letter α designate the chosen level of significance, possibly $\alpha = 0.10$, or $\alpha = 0.05$, or $\alpha = 0.01$, etc. Further, let ξ designate the adopted conventional measure of the effect of seeding, and $\beta(\alpha, \xi)$ the corresponding power function of the statistical test to be used.

The category of tests we use routinely is that labeled the optimal $C(\alpha)$ tests, developed in Neyman (1959). The advantage of these tests is that they are easily applicable to testing composite hypotheses of unconventional type, such as are frequently encountered in real life. Their disadvantages include the fact that their attractive properties are only "asymptotical" and "local." Thus, if N designates the number of observations, say $N = 50$ or $N = 100$, etc., Monte Carlo simulation is needed to see whether the asymtotic properties of the tests are satisfactorily approximated or not. In our experience, $N = 50$ is frequently sufficient.

A variety of relevant questions are discussed in Neyman and Scott (1967b). One of these is how to define the conventional measure ξ of the effect of cloud seeding. Here, both the conceptual aspect and the simplicity of methodology must be taken into account. For example, it is certainly desirable to discover whether seeding affects the probability θ of the experimental day being "wet." How does one define ξ? One possibility is to let ξ denote the difference between the probabilities of some rain in the target with and without seeding. However, considerations of convenience in applied work brought us to adopt, say $\xi_1 = (\theta_s - \theta_c)/\theta_c$, where the subscripts s and c connote "seeded" and "control," respectively.

In addition to discovering whether cloud seeding affects the probability of some rain in the target, it is also important to find whether it affects the rainfall per "wet day." Finally, there is also the consumer oriented question of whether the seeding of clouds affects the target precipitation averaged per "experimental day," whether wet or dry. All this is the subject of a paper by Neyman and Scott (1967b). It provides three test criteria, labeled Z_1, Z_2 and Z_3.

Criteria Z_1 and Z_2 are optional $C(\alpha)$ tests of hypotheses H_1 and H_2, that cloud seeding does not affect the frequency of wet days and that it does not affect the rainfall per wet day, respectively. The criterion Z_3 is not a $C(\alpha)$ test but a palliative oriented toward cases in which the two mechanisms M_1 and M_2 work in the same direction. It is a linear combination of Z_1 and Z_2 so adjusted as to maximize the chances of detecting the falsehood of the hypothesis H_3 that the seeding does not affect the target precipitation averaged per experimental day, whether wet or dry. The general idea is that, while the effects of mechanisms M_1 and M_2 may be weak and, thus, difficult to detect, the combined effect of their working in the same direction may be noticeable.

Table I illustrates the use of the three test criteria Z_1, Z_2 and Z_3. It refers to our evaluation of the National Hail Research Experiment (NHRE). The purpose of this experiment was to find whether the seeding of clouds on appropriately defined "suitable" days could diminish the hailfall. The experiment continued over three summers in 1972, 1973 and 1974. It happened that in 1973 a mishap occurred "resulting from an effective error in radar calibration that biased the sample toward more vigorous storms." (NHRE, 1976, p. A.2) There were only six experimental days in 1973.

Table I Preliminary evaluation of adjusted hail mass, NHRE 1976a separator data. Effect of seeding on frequency of hail days, on mass of hail per hail day and on mass of hail per experimental day.

Years studied	Seeding	Frequency				Amount per hail day			Amount per experimental day		
		No. of zero	No. of hail	Effect	P	Mean	Effect	P	Mean	Effect	P
All 3 years	S	10	17	+5	1.00	2901	+14	0.80	1826	+19	0.75
	NS	12	18			2550			1530		
1972 + 1974	S	9	16	+19	0.65	2958	+160	0.12	1893	+209	0.088
	NS	12	14			1137			612		

In consequence of this information, we performed two evaluations: one, including data of all three years and the other based on data from the two "unbiased years." Table I reproduces the results we obtained. Looking at the part of Table I referring to the "unbiased years," it is seen that with seeding the frequency of some hail and also the amount of hail per a "hail day" were greater than without seeding. Yet the "percent effects" were too small to approach the customary standard of significance. However, the third multiple column of the table summarizes the application of the palliative test criterion and thus gives the combined effect of two mechanisms M_1 and M_2 working in the same diretction. The result is quite suggestive that the basic idea of the effectiveness of seeding at the NHRE deserves some revisions.

As mentioned, the palliative test criterion Z_3 is not the optimal $C(\alpha)$ criterion to test the hypothesis H_3 that seeding does not affect the target precipitation averaged per experimental day, whether wet or dry. Work to deduce such a criterion is now in progress on a impressively international level, including scholars in Brazil and in Ghana! I am looking forward to seeing the results.

5 DANGER OF WRONG CONCLUSIONS WHEN THE TWO MECHANISMS ARE IGNORED

The relevance of the two mechanisms M_1 and M_2 seems to have escaped the attention of many practitioners of cloud seeding. Quite frequently, their evaluations of the randomized experiments are based on the Mann-Whitney U-statistic. This test detects the falsehood of the "null hypothesis" that two compared distributions are identical. I am indebted to Paul C. C. Wang, for the preparation of Figure 2. It gives plots of two cumulative distributions of nonnegative random variables, distributions that are rather different. With enough observations, the Mann-Whitney test criterion is sure to discover that these distributions are not identical. If one of these distributions is that of the target rainfall with seeding and the other without, the rainfall averaged per experimental day wet or dry, the conclusion indicated by the U-test would be that the seeding had a real effect. Yet, the two distributions in Figure 2, reflecting the mechanisms M_1 and M_2 working in "opposite directions" are so adjusted as to imply no effect of seeding per experimental day whatever!

6 CONCEPT OF VARIABILITY OF RESPONSE TO CLOUD SEEDING

The concept of variability of response to cloud seeding appears to be due to E. J. Smith (1967) from CSIRO, Australia. Having developed the suspicion, Smith called for the development of an efficient statistical methodology for the detection of the phenomenon if such exists.

The question raised by Smith is quite important. If the study of a particular experiment reveals the presence of variability of response to cloud seeding, this would affect the interpretation of the average apparent effect and would stimulate the atmospheric physicists to identify the weather conditions at which the response was lower than the average and also those at which it was higher. One illustration must suffice.

Currently there are indications that the effect of seeding convective clouds of the air-mass type is different from that of frontal types. If the original solution of a relevant experiment included the routine of using a really effective test advocated by Smith, the question of the air-mass/frontal difference might have been asked, and perhaps answered, long ago.

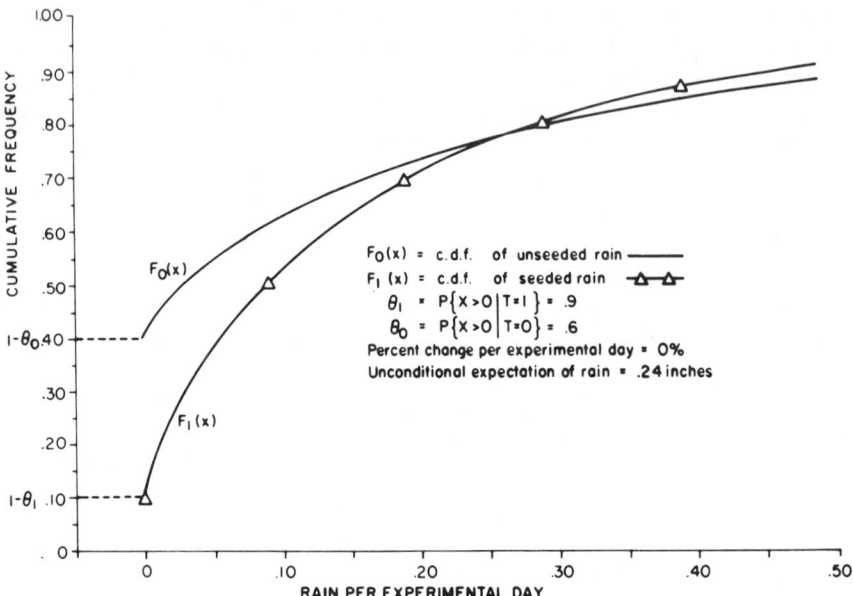

Figure 2 Evaluations of cloud seeding experiments that ignore the two mechanisms of the effects may lead to false conclusions. The two exhibited distribtion functions that could correspond to days with and without seeding are rather different. Yet they reflect the case of no effect of seeding on precipitation averaged per experimental day, whether wet or dry.

A response to Smith's call for a test for homogeneity of response to treatment (whether cloud seeding, or some medical treatment) has been offered in Neyman and Scott (1967a) as one of the examples of the use of the optimal $C(\alpha)$ methodology. The applications to the detection of variable response to cloud seeding are limited to rainfall per wet day.

Subsequent developments involving several authors have been recently reviewed by Paul C. C. Wang (unpublished). Formula (2), taken from Wang's work, gives the optimal $C(\alpha)$ criterion, for the detection of the variable response in cases where the distribution of the nonzero precipitation has a gamma density as described in the earlier sections of the present article.

$$\hat{Z}_N = \frac{\hat{\gamma} N_t}{\sqrt{N}\,\hat{\sigma}} \left[\frac{\hat{\gamma} \Sigma_t (x - \bar{x}_t)^2}{N_t \bar{x}_t^2} - 1 \right] \qquad (2)$$

Here, the subscript t refers to seeded experimental days and $\hat{\gamma}$ denotes the maximum likelihood estimate of the shape parameter obtained using all the N observations on the assumption that there is no variable effect of seeding.

Heuristically, it will be seen that the criterion (2) will tend to indicate the presence of variable response when the sample variance of seed day precipitation is, in a sense, too large.

Wang's comments include the indication of an oversight in my original developments that were put right by LeCam (1960).

7 CONCLUDING REMARKS

As mentioned at the outset, the most sophisticated mathematical concept generated by studies of meteorology appears to be the concept of characteristic functional. It was reintroduced by LeCam in his Note of 1947, even though it was first created by A. N. Kolmogorov and published twelve years earlier. The subsequent concepts described in this article are indeed "mathematically pedestrian." However, from the point of view of mathematical statisticians who contemplate studies of weather modification, a familiarity with them is likely to be important.

Acknowledgments

Research for this report was supported, in part, by Department of the Army Grant DA AG 29 76 G 0167 and Office of Naval Research Contract ONR N000 14 75 C 0159. All opinions expressed are those of the author.

BIBLIOGRAPHY

Ferguson, T. S. (1961). On the rejection of outliers. *Proc. 4th Berkeley Symp. Math. Statist. Prob.* Los Angeles & Berkeley: Univ. of California Press 1, 253-287.

Green, R. F. (1974). A Note on outlier-prone families of distributions. *Ann. Statist.* 2, 1293-1295.

Green, R. F. (1976). Outlier-prone and Outlier-resistant Distributions. *J. Am. Statist. Assoc.* 71, 502-503.

Grubbs, F. E. (1969). Procedures for detecting outlying observations in samples. *Technometrics* 11, 1-21.

Harris, T. E. (1963). *The Theory of Branching Processes.* Berlin, Göttingen, Heidelberg: Springer-Verlag.

LeCam, L. M. (1947). Un instrument D'étude des fonctions aléatoires: La fonctionnelle charactéristique. *C.R. Acad. Sci. Paris* 224, 710-711.

LeCam, L. M. (1960). Locally asymptotically normal families of distribution. *Univ. of California Publications in Stat.* 3, 37-98.

NHRE Staff (1976). Revised Plan for the National Hail Research Experiment. National Center for Atmospheric Research, Boulder, Colorado.

Neyman, J. (1959). Optimal asymptotic tests of composite statistical hypotheses. *Prob. and Statist.* (The Harald Cramer Volume). Almquist and Wiksells: Uppsala, Sweden. 213-234.

Neyman, J. and Scott, E. L. (1967b). Note on techniques of evaluation of single rain stimulation experiments. *Proc. 5th Berkeley Symp. Math. Statist. Prob.* Los Angeles & Berkeley: Univ. of California Press 5, 371-384.

Neyman, J. and Scott, E. L. (1971). Outlier proneness of phenomena and of related distributions. *Optimizing Methods in Statistics*, New York and London: Academic Press, Inc., 413-430.

Smith, E. J. (1967). Cloud seeding experiments in Australia. *Proc. 5th Berkeley Symp. Math. Statist. Prob.* Los Angeles & Berkeley: Univ. of California Press 5, 161-175.

Some Operational Considerations in Evaluation Of Weather Modification Programs: A Short Excursion into Epistemology

Pierre Saint-Amand

Research Department
Naval Weapons Center
China Lake, California

¿UNDE ID SCIS?

The history of science is an essential component of the education of anyone expecting to work in science or expecting to be able to evaluate the work of other scientists. The most important lessons to be learned do not derive from the content of the discoveries but from the methodology used by the various scientists. Science grows as a result of the combined efforts of many different persons. The techniques used by these persons are as different as the people who use them. The techniques work to the satisfaction of those who use them, but not necessarily to the satisfaction of others schooled in different disciplines. What may seem as satisfactory evidence to one may not be acceptable to others. Indeed, knowing the right answer to a problem does not necessarily constitute proof. Proof is elusive at best. What passes for proof or for truth is usually a demonstration of self consistency within the paradigm employed by the individual or by his school.

The roles played by scientists are quite different depending upon the functions that the person performs or is expected to perform. Innovators reach conclusions, often through methods they do not understand or do not even know they are using. These people are usually loathed during their lifetimes by project managers and by scientists operating at a different level. Engineers usually regard them at best as uneconomical dreamers, at worst as accidents waiting to happen. In order to support the facts that they have discovered, the innovative scientists, and this includes mathematicians, often construct intellectual artifacts of dazzling perfection to justify what was originally a little more than a dimly perceived idea, or perish forbid, a concept arrived at by pure empiricism. The rigorous proof developed to demonstrate that some theorem is true is an example. Fermat's last theorem is probably quite correct, yet no proof exists for it. I do not believe that Fermat had a proof, even though he wrote in the margin of his book that he had a marvelous proof that the margin was too small to contain. If such a proof existed someone should have produced it by now, considering the amount of time and energy that has been expended on such a simple appearing concept.

The number of people who produce new ideas is not large, but those who contend they do are so numerous that it is difficult to tell if anything has been produced, worthwhile or not. In order to be sure that they have done so, it is necessary to test these ideas. Many ways lead to satisfactory testing and some are more productive than others—the exact choice depends upon the style of the investigator and the nature of the problem. You gentlemen are apparently appointed to be the tester of ideas. Still others are anointed as the recipients and preservers of the knowledge generated and screened by the process. These latter are learned men, it is true, but by nature are so selective and rigorous that they are usually not qualified to recognize, to discover or to develop new ideas. Some notable exceptions exist, but these are truly exceptions and usually were welcomed into the academic community after having attained some distinction outside.

One of the more fundamental failings of the educational system in teaching people to be scientists is that the wrong tendencies are reinforced. Perhaps I should say overemphasized, overemphasized to the exclusion of the development of the sort of personality that can discover things. From childhood on, an individual is reinforced for reinforcing the sense of self righteousness of his elders, his neighbors and his peers. By the time he is six, his schools have precious little left to work with and they meticulously perpetuate the same tendencies. The church does exactly the same. This sort of behavior has survival value and even leads to domestic tranquility so long as the individuals are exposed to minimal threats and no serious thinking or leadership is required from them. The very fact that the system has survival value has encouraged and generated it.

Rewards, within the education system, come, not from learning truth, or even self-consistency, they come from being able to recreate what the professor has said in such a way that his self-confidence is not challenged. The risk of appearing stupid because one has reached a conclusion at variance with ideas widely held is a great deterrent to clear thinking in a culture that reinforces intellectual conformity. I recall a professor who flunked several of his students for recognizing an error in one of his derivations. I also recall Beno Gutenberg deliberately making an error in a complicated derivation and then flunking those students who didn't find it. He was truly exceptional, however, and his like is rarely encountered on the academic scene.

When I was fortunate enough to teach college, many years ago, I experimented with ways of making people more productive of ideas and more critical of what they were taught. The result was a disaster insofar as the rest of the faculty was concerned. The students loved it and although their general grades dropped a little at first they soon recuperated. Out of some 80 undergraduate students in to the first two graduating classes, some 12 have gone on to the PhD degree and perhaps 20 others have obtained masters degrees or undertaken graduate work. I still hear from them and have watched their progress with interest. Having been exposed to the same treatment myself from Beno Gutenberg, Hugo Benioff, John Buwalda, Robert Sharp, and other faculty members at Cal Tech; I know what they went through.

I have worked in research and in the application of new knowledge to human affairs. Although I have not personally accomplished much, I have been able to observe some of the most innovative minds in the country and to some extent have been able to figure out the processes they used. The processes are quite deliberate, even though the individuals do not appear to understand them and would not, or could not, tell you what they were. They have some things in common, they do not become enamoured of one idea, they are quite critical of themselves and others, they do not choose to believe what the mass of their peers believe unless they have looked into it themselves. They are not afraid of failure, nor of appearing stupid. They are risk takers. They ignore details and they usually can tell what is important to pursue and what is not. They treat what they believe is not important in a very cavalier way. They are not perceived by their colleagues as having common sense or even very good sense. They, moreover, have a strong emotional bias towards believing that they can do the things they set out to. This latter point is the keystone of the bulwark of disagreement with the academics.

If you want to accomplish a particular thing you will do better to seek help from one who believes it can be done than from one who doesn't. I recall Prof. Fritz Zwicky saying that anything that man can imagine he can do, and he probably will do, provided it does not violate the second and third laws of thermodynamics. Zwicky's attitude seems a little extreme, even to me, and his colleagues, justifiably distrustful of a flame that burned so brightly, were eternally suspicious of his zeal, even when he was right—as he usually was.

The reason for this rambling discourse is to set the scene a little. People who do applied research differ from those who do regular research. People who teach are different still. Each could learn from the other. Usually, however, this does not occur. Researchers rarely know what they are going to do very long in advance; they enjoy uncertainty. They are not necessarily orderly. Project managers do not enjoy uncertainty, they must be assured in advance that something specific will be done. Moreover, they often cannot tell, or have no way to tell, if it has been done or not. A special case of this type is the modern "executive" who, by use of "management by objective," has totally avoided the problem of being faced with cognitive dissonance arising from the technical content of their work. Teachers do not like uncertainty at all and cannot abide cognitive dissonance. Researchers thrive on it.

The one common thing that most of these people have is a lack of time, tranquility or inclination to study a problem that they are working on. Having been assigned a goal, they proceed to do what they can without looking far enough to see if it makes sense to do it, because they have learned to please rather than effectively to serve.

An example: If a rain dancer were to materialize and announce that he would make rain, most statisticians would not trouble to set up a controlled experiment to see if he did indeed make rain. On the other hand, a good many otherwise sensible people are now trying to see if animal behavior is a precursor to earthquakes because someone said that he predicted an earthquake on the basis of animal behavior (Chu, 1976 and Raleigh, 1976). Actually, the prediction was probably not made, and a good many were made that didn't come true, but nonetheless they did indeed work the subject over (U.S.G.S., 1976). If a professor at a well-known university were to conduct an experiment in weather modification you may be assured that considerable attention would be given to his results and very high powered, sophisticated reasoning would be brought to bear on the matter.

This has happened. The sad thing is that the people making the analysis did not trouble to find out if the professors knew any more about rainmaking than did the dancer. The conclusion to be reached from such an experiment should never be extended to embrace other efforts. Indeed had they thought the experiment through and acted as mature scholars rather as learned technicians, they would have concluded that the main thing that was proved was that those particular experimenters did not seed clouds effectively.

Irving Langmuir was a very innovative individual. Within a very few years of the end of World War II, in which he had been involved in the development of screening smokes and became interested in cloud physics, he thought of and tried most of the different techniques that have been used in weather modification. All of his cerebration is written out in his notebooks (Suits, 1961). He deduced most of the thoughts from recourse to physical fundamentals. His reasoning is as correct as one could wish, considering the state of the art when he wrote it. He was of course ably and imaginatively assisted by Bernard Vonnegut and Vincent Schaefer. If one were to attempt to organize a new program in weather modification he would be well advised to read these notebooks. Little that is new has been thought up since his time. Most of the techniques that we tried at the Naval Weapons Center came directly from these works. I am sure that Langmuir and his colleagues knew that what they had done was right. They were hardly able to prove any of it to the satisfaction of the community at large. Even when it came to clearing supercooled clouds and were writing their initials in a cloud deck, they were taken to task because it might have "happened naturally." Knowing you are right and having proof are two different things; convincing others is still a third. Almost thirty years have elapsed and we are still faced with the same problem. There is no easy way to settle it.

Enough of this sort of talk. Let us see now what is needed to perform an experiment to tell if one has modified the environment. One can proceed in several ways and in each way reach satisfactory conclusions if some thought is applied. The first way to proceed is to try to learn about the processes to be employed, and to check at each step as many physical parameters as possible to see if the changes that were predicted are taking place in any case. The second way is to treat the whole matter as if no one knew anything about the subject and to observe only the output of an experiment without any understanding of the processes. This latter method leads to continued scientific study and to no useful conclusions. A variant of this technique is to set up an hypothesis and then to test that hypothesis. An infinite number of wrong hypotheses can be set forth and systematically tested. Indeed this process is popular in some circles wherein hypotheses that were at variance with the laws of thermodynamics, and which could have been dismissed out of hand, were tested at great length using statistical methods that could not have answered the question in the first place because of lack of an adequate sample size. This method assures protracted support at high levels and only rarely produces useful results. We will return to this method of testing in a moment.

The first method is that ordinarily used in science. One conducts experiments trying to control variables as closely as possible so that he can observe the effects of small changes in one variable upon the outcome of the experiment. Having good theoretical reason to expect that certain changes will take place, he looks for and measures them—systematically exploring the attainable range of the variable. He then does the same with another, and then another, until all the ones he can think of are exhausted. He then tries them in combinations and eventually arrives at an understanding of the phenomena he has been working with. This understanding is empirical, and he ought not use it beyond the bounds of the experiment. The conscientious researcher will, at this stage, use statistical methods to evaluate just how sure he is of each relationship. There it rests unless some flash of insight illuminates the corners of his mind and he can see that what he has done is a part of a larger concept and successfully defines this concept. This is the stage to which all hope to arrive. Usually what happens is that the starting theory is wrong in some way and that divergence between predicted and observed facts leads to a new postulation, often totally different—and usually pointed out by someone else. This is the method used by physicists. It works well for them because with consummate skill they can control the factors involved.

It does not work for some others, however. Astrophysicists cannot control the variables in their experiments; they cannot manipulate the stars. They therefore have to devise observational programs in such a way that they can observe a plurality of objects having a wide range of properties and select from these the range of variables that they need to see if such things are related. This process is powerful and the lack of understanding of certain critical issues exposed by these methods has led to most of what we know about radiation theory. It is fraught with danger that any statistician will point out: that a relationship observed between two or more variables may not mean that the variables are related. This method is used by upper atmospheric physicists. It is virtually untouched by those who deal in the lower atmosphere.

Geologists are in a different class. They can usually only observe a very small portion of what they must see to understand their problems. They cannot usually make experiments. To the extent that this is possible they do so, but in view of the total science they don't have much leeway. They use careful observation, indeed, they develop observational qualities to the point where they can distinguish tiny differences that maybe don't even exist. They use a method of thinking called the method of multiple hypotheses. This is also a powerful tool, and finds considerable application in intelligence analysis and other arcane branches of intellectual endeavor. It usually leads to an acceptably correct solution, although it does not really generate proof, unless convergent tests are developed to test the validity of the solution.

Electrical engineers, computer experts, and similar folk, who are really applied mathematicians, use a different method. They regard certain affairs as black boxes. They put step functions into the system under question and observe the indicial response of these systems. Knowing this they can predict the response of the system to that stimulus. They can never know for certain the inner nature of the machine by a set of such observations but they can tell you how it will function for any input of that particular variable. This method is used on large scale by psychologists, intelligence experts, political manipulators, and businessmen to determine what people are thinking and how they will react. It's a chancy business but in the proper hands it works well.

Legal proof is quite different. That profession uses a strictly Euclidean logic. This is difficult because the subject matter does not readily fall into Euclidean dichotomies without subdividing the statements into a whole series of lesser statements that are either true or untrue. This is often not feasible. Rather than arrive at truth or justice the courts seek an acceptably satisfactory solution within the totality of the circumstances. They make a statement and then try to show that it is true or false. This is done by choosing two champions to set forth the best points, or at least those points that appear best, of the argument they have chosen to try. Evidence is brought forth by each side to support their points of view. Evidence to the contrary is not brought forth except by the other side. According to the rules of the game it is not necessary to bring up material deleterious to your argument. It is wise to have such material well in mind, however, because if any reference is made to subject matter touching upon the occluded material it may be brought out upon cross examination. The degree of illumination given to poorly phrased, or badly understood, components of the argument and to drawing distinctions between almost identical statements is frightening and it may be said that with competent lawyers the degree of rigor in proving each point is greatly in excess of what a scientist would regard as an uneconomical overkill. The factual matter is not only explored, any possibility of bias is ruthlessly exposed, the characters and reliability of the witnesses are scrutinized. From that point on, the system is not quite so rigorous because a jury usually of people who do not have much else to do, or of a judge who may or may not be of high intellectual quality, then decides. This latter aspect is ameliorated somewhat by an appeal process. A scientist usually suffers from cultural shock after exposure to a court room. Most lawyers do not find life in a laboratory the least bit enlightening. It has been seriously proposed, most recently by Arthur Kantrowitz, that some sort of a quasilegal system be tried to test the validity of competing scientific arguments and a conference attended by some 200 persons was held in 1976 to discuss the idea. It would be interesting to try—except that the judges would probably be chosen from the hard core establishment. I suspect that if the method were in widespread use, that we would still be Aristotelians.

Now then, one wishes to do research in, say, rainmaking. It is well known that water will neither condense nor freeze unless it has something to condense upon, or unless something causes it to freeze. Water has been cooled to at least 100°C below zero in the laboratory without having it freeze. A variety of solids, indeed, any solid at all, will cause water to freeze at some temperature provided that the time required for it to dissolve is greater than that required for the water to freeze. The temperature at which it will be caused to freeze is called the activation temperature of the material; it depends upon the surface free energy of the substance, and this in turn depends upon the distribution of atoms and molecules upon the surface of the material. If the activation temperature and solubility of a substance is known, all of its properties as a nucleant can be predicted. This theory, called the theory of catalysis, has been worked out in great detail by the physical chemists. The theory predicts quite correctly that unstrained silver iodide will cause water to freeze at $-2.3°C$ (Turbull and Vonnegut, 1952).

We can experiment with the weather by causing water to condense upon something, or we can cause it to freeze. Let's stick to the freezing concept because we have some other information that may be useful. Based upon impeccable physics, we know that frozen water has a lower vapor pressure than liquid water at all temperatures below the triple point where ice, liquid and vapor, can coexist. Thus, if we consider putting some material into a cloud that should cause water to freeze, we can maybe increase the rate of condensation onto the frozen particles. If we can do this we can expect them to grow at the expense of their neighbors who may even be caused to evaporate because the vapor pressure of the ambient air has been lowered by the newly formed snow crystals.

We know that droplets and snowflakes fall if they are large enough and heavy enough. We know intuitively that if some of them are falling faster than their fellows they may collide with others, and if they stick together, we might expect them to grow—indeed, grow large enough to become quite efficient at scavenging other drops. We also know that when water condenses and/or freezes it gives off heat. We know that when air rises it expands and cools and hence we should be able to expect that we can modify clouds with some success—provided we have any faith at all in the laws of physics, chemistry and logic and actually believe that they apply in the atmosphere. There is little reason to doubt this although some meteorologists do not seem to believe it.

What is now incumbent upon us is to ensure that we will seed these clouds and that what we have set forth is indeed true. All of a sudden things get a little harder. We might set out to calculate the results first, as any good computer operator would. To do this we would have to make all sorts of assumptions that we could not test. We would probably parameterize the parts we didn't understand and then truncate the whole program in each of several dimensions. Worse yet, we might need or make some assumptions that we didn't know about. For a number of years this practice was quite prevalent in weather modification. One technique was to make a model and then to "tune" the model by causing it to conform to observations made to test the theoretical model. Often several tens of variables, mostly parameterized, are involved and this procedure soon leads to a quite satisfactory fit over the range of observations for one particular cloud and yet may be completely wrong. One such model had as a basic premise that droplets grew in the cloud until they reached a certain size and then split forming two droplets, each of which grew. This process, calculations showed, should repeat itself every thirty seconds or so. For short runs of the model it worked well, but real clouds often last for hours, or even days. Had the model been used to predict factors much beyond the regime calculated it would have been necessary to import water from elsewhere in the universe to keep the cloud going. This effort was mercifully dropped. We could spend the rest of our lives in this state of mental masturbation without results. We must identify and verify the fundamental physics, chemistry and maybe mathematics first.

We therefore elect to be a little more explicit and try to verify the assumptions in the preceding paragraph. No trouble with the activation temperature of silver iodide—we can verify this experimentally (Cuilong, 1947; Vonnegut 1949; Schaefer, 1949; Aufmkampe and Weikmann, 1951; Odencrantz, 1969). We can measure the solubility of silver iodide (St. Amand, Mathews, Reed, Burkardt and Finnegan 1971). We can calculate the time it takes to dissolve (Mathews, Reed, St. Amand, and Stirton, 1972) in a droplet of given size and temperature. We can use the catalysis theory to estimate how big we should make our particles if we expect them to function, we can calculate the temperature at which they will function. Right away we find out that we cannot produce as many particles as we might like on the basis of economy because they will not work at a reasonably high temperature and moreover they will dissolve in the droplets. We therefore avoid one of the pitfalls that early experimenters had fallen into by increasing the particle size over and above what they had used. We can thus convince ourselves that the nucleant will indeed cause water to freeze. We also need to know if the nucleant will work by sublimation, condensation followed by freezing or by contact nucleation. This took a long time to decide. It turns out from study of Brownian Diffusion, Smoluchowskian capture, diffusiophoresis and electrophoresis that it doesn't really matter because the particles will be captured by droplets as soon as they grow to a reasonable size, provided we use enough particles. Having calculated the capture kernels for particles with drops we decide that it doesn't make much difference how many we use if we can obtain a volume density of about 10^2 per cubic cm. This is easy to do over a small volume.

The next thing we need to know is if the droplets will freeze, grow, and fall. This can be calculated by a variety of techniques. As to whether the drops will collect each other or not, we have a thorny time deciding because the so-called scientific community was convinced until very recently that droplets smaller than 18 microns could not be captured by other larger droplets, or if they were it was such an infrequent event that it didn't much matter. This opinion derived from an exercise in the use of an electronic computer by Irving Langmuir (1948), who was just learning how to use a computer, and while trying to work out capture cross sections for droplets employed, as I understand it, an over simplification of his basic theory. Hocking (1959) soon recalculated the affair and made a similar mistake. The ficticious barrier to capture came to be called the "Hocking Cut-Off." It was apparently an artifact of the computational technique. Nonetheless it was cited in many learned quarters as evidence that tropical clouds must reach above the freezing level because they could not rain by a collision-collection process alone. A number of workers at NCAR and UCLA finally resolved this point and Mathews recalculated all the collection coefficients by going back to first principles, eliminating the simplifying assumptions. Work by Berry and by Gillespie then proved conclusively that capture took place often enough to be an important factor in cloud development and that it could indeed rain. Having thus satisfied ourselves that the main basis of our scheme was well founded in physical theory, it is possible to see that an unbroken chain of logic exists from the placement of material in the cloud to the fall of water. Mr. Elliott has worked this logic chain into a good computer model for orographic clouds. Why then should it prove difficult to establish whether or not one can augment rainfall?

It is not easy to do this even though the physics tells us that it will work. The reason is that it is easy to go astray unless you think your way through the problem. We might consider the following:

A. Weather modification should work if:
 1. One seeds clouds that can be frozen.
 2. One uses nucleants that induce freezing under the conditions in the cloud.
 3. One places them correctly.
 4. One places them at the right time.
 5. One allows enough time for them to work.

B. The results will depend upon a few things we did not discuss yet:
 1. You must do it right.
 2. The temperature and humidity profiles.
 3. Wind shear.
 4. Wind field.
 5. Degree of activity of natural nuclei, size spectrum and activation temperature.

C. If we are to test it we must, in order to be effective, show that:
 1. All of the above have been considered. That all the things in Section A have been done correctly.
 2. That any comparison areas, clouds or times be truly independent.
 3. That the weather is the same, or nearly so, at the times and places used for comparison.
 4. That enough samples be taken to overcome significant vagaries of the weather.
 5. That enough samples be taken to permit stratification by a number of variables and still retain meaningfully large samples for each stratification. This means stratification by temperature profile, by humidity profile, by wind shear profile, by wind speed, by wind direction, by general synoptic weather patterns, etc.

No sooner was it decided that cloud seeding might be promising, than a number of experimenters began trying to prove that it was. They began trying to prove it before they began to understand what they were doing. An enormous amount of research was expended trying to see if previously selected schemes worked. Very little was done to try to find out how.

The commerical operators did better than the researchers because they were trying to optimize the results rather than prove an hypothesis and they employed a more organic approach. In any case, it serves to recount some of the misadventures that befell the experimenters because they did not carry out all the steps in the chain.

I have a copy of a report prepared by a French hydraulic engineer in 1958 covering a four-year experiment conducted in the Haute Isere of France by North American Water Development Corporation, for Electricite de France (Serra, 1958). See Figure 1 for a schematic representation of this experiment. This experiment is unique for being well thought out and well conducted. Monsieur Serra was not a meteorologist, nor was he a mathematician. He did, however, understand well how to go about such an affair. Dr. Krick's people did the seeding with a device that produced silver iodide. The seeding was done on a storm scale by turning on a series of burners in such a way that the wind would blow the combustion products into the target area, an elliptical zone about 30 kms across. The seeding was conducted on every opportunity during the course of four years. The test consisted of establishing a correlation coefficient with the rainfall in other areas to that in the target area for a long period preceding the experiment. It consisted of comparing rainfall, run-off and the shape of areas of anomalous rainfall. Serra's work is extremely well done and quite rigorous, especially considering the day when it was prepared. Krick's part of the work was well done because he used a good seeding agent and was quite expert at the targeting based on wind patterns. Not everyone can do this; Mr. Robert Elliott is one of the very few who knows how. A weakness of the experiment is that some of the control area was probably seeded from time to time, thus diluting the apparent results by making the control areas look a little better. We have since learned that increases in rainfall occur upwind of the seeding site, presumably brought about by increased convection. To some extent, Serra's effort suffered from this defect although he didn't know it. But again, this defect would have only made the results seem poorer than they actually were. He violated B2, C2 and possibly B5 but was still successful.

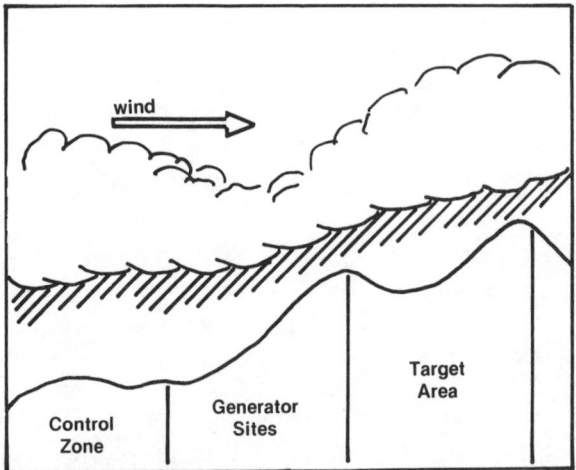

Figure 1 Layout of Project Tignes. The whole landscape slopes gently uphill. If seeding induces growth of clouds over the target area the upwind area should have a general increase in wind speed. This will increase orographic uplift over the upwind (control area) and cause an increased rainfall there as well as in the target area.

A similar seeding experiment in Utah was scuttled by an investigator who used information only from rain gauges in an area not really affected by the seeding at the time it was done. He ignored snowfall and stream runoffs, even though these had increased in the seeded area as was planned. So thorough was his analysis that it was some years before seeding was resumed in Utah. He violated most of the above. The state is now using weather modification very extensively.

Two massive experiments were performed subsequently, that have attracted a great deal more attention than they deserve and these have served as the basis for an overall criticism of weather modification in general. The project called White Top has contributed more to the confusion than any other single effort. This experiment was conducted with a great deal of rigor. The results have been discussed endlessly. It was a well done experiment and the background and reputation of the organization and the persons were impeccable. Thus those investigating this effort did not look beyond the bare facts of the results. What got the Chicago group into trouble was that they did not really seed any clouds at all, or if they did so it was not in the way they intended and probably a counter productive way—as Mr. Elliott will explain later when he talks about cumulus clouds.

In 1949 Vonnegut published a paper on the use of silver iodide as an ice nucleant and described a whole series of ways of making silver iodide smoke; most of them worked and did indeed produce silver iodide. He then published a paper a year later in which he told how to make acetone burners and how to cause silver iodide to dissolve in acetone. Silver iodide is poorly soluble in most anything. In order to get it to dissolve one must complex the silver iodide with another compound so that the complex molecule is soluble but not dissociated. This permits large quantities of the material to go into solution without exceeding the solubility produce constant of the silver iodide. One can use ammonium iodide or an alkali iodide to promote the solution; either works equally well for the purpose. The solution is then placed in a tank and expelled through a nozzle and burned. The burning destroys the acetone and presumably leaves a little ball, of whatever was dissolved originally, as an aerosol particle. When ammonium iodide is used, the ammonium iodide is pyrolysed, releasing free iodide and silver iodide. When sodium or potassium iodide is used, the alkali iodide remains intact, it being more thermodynamically stable than the silver iodide. For some reason the entire weather modification community began to use acetone burners and to use the alkali iodide-silver iodide complexes instead of the very satisfactory ammonium iodide solution.

This might seem like a little thing, but the effluvia from the alkali iodide-silver iodide solutions is water soluble to a high degree. In fact, it is so deliquescent that it will take up water from the air and dissolve itself, at relative humidities of the order of 30 to 40 percent. Thus, used in the vicinity of clouds, but not directly within the cloud, the material would dissolve shortly after being emitted. Dr. Booker once attempted to trace a plume from an acetone burner that was using an alkali iodide-silver iodide complex. He was unable to do so for more than a very short distance because the material dissolved in the humid air. Vonnegut realized that this was a serious problem and in the Report to the President on Weather Modification, published in 1957

(Vonnegut, 1957), admonished workers in the field to be aware of this distinction. Project White Top used an alkali iodide-silver iodide complex, released not directly in cloud but at some distance on the outside. The material was totally ineffective in the manner in which it was intended to operate.

Similarly a project conducted at the University of Arizona used a similar solution emitted some miles from the clouds, with exactly the same results. Nothing definitive happened. Both groups of experimenters had violated items A 1, 2, 3, and 4 above. One should not have expected any results. Moreover, in a process described in some detail in 1883, the series of reactions of the alkali iodide-silver iodide complexes is in itself very complicated. It is quite possible that in Project White Top, the droplets of alkali iodide-silver iodide solution became dilute enough to permit the complex to dissociate and the silver iodide to be precipitated in colloidally sized particles that would only work at temperatures about -14 degrees C or colder. This is no help at all because more than enough nuclei exist in ordinary air to nucleate the whole top of the cloud at that temperature. This is well above where the extra nucleation will do any good and where it might even cause some local diminution of rainfall. If any precipitation is induced, it might be due to secondary seeding some miles downwind in the direction of the upper level winds just as Dr. Neyman has shown from his analysis.

An organic approach to the problem would have gone somewhat differently. A less rigorous, less detailed explorational experiment would have been done first. Material would have been placed directly into a cloud. It would have been possible to tell if freezing took place or not. No great amount of equipment is required for this. A black glove held out the window of an aircraft would have done the trick and one could have told if the freezing nucleant worked. It is easily possible to perform the experiment without an aircraft. The appearances of the clouds change. Certain optical effects occur in the presence of ice and not in the presence of water. Crystals could have been collected and counted. The people at the University of Wyoming do this in their cloud top laboratory at Elk Mountain. They have sight tested many nucleating compounds and are able to tell immediately if they work. Having decided if the seeding agent worked, the next step would be to try it on a cloud on a larger scale, not in the air outside the cloud but directly in the cloud. This was done by Langmuir and his colleagues. He had cleared off large areas of supercooled cloud and had seeded cumulus.

Seeing that the material fell into the cloud is important because, if placed directly into the cloud mass at heights greater than that at which the activation temperature is reached, even the alkali iodides work reasonably well. Dr. Bowen's experiments in Australia were successful, even with an alkali iodide, because he insisted that the clouds be seeded exactly where it would do some good and not outside the clouds. He even noted the $-14°C$ level of activity. My point is that one does the experiment so that he knows that he has performed the manipulation he intended. If the experiment does not work then he tries something else. If it does work, he then tries variants on the theme to get a feel for the situation. Having thus satisfied himself, one point at a time, the investigator goes on to check the next.

It soon becomes possible to home in on a useful method and to develop the techniques necessary for making the interim decisions as to whether a thing was being done right or not and for deciding how much should be done. Enough physical changes can be observed, usually easily and cheaply, to chart one's course clearly to the development of a useful technology. This technique is commonly used by inventors and innovators in general. It is also used by people who like to manipulate others. It is the same technique used by B. F. Skinner and Kellar Breland to teach chickens to distinquish between geometrical figures.

Now then, after having defined the problem and calibrated the tools and decided what one can do, it is proper to seek a quantitative answer to the question "Does my technique produce detectable amounts of rain?" We then answer the question "how much?" and so on. It is not necessary to perform a large statistical experiment to reach a satisfactory decision although it is definitely an elegant and satisfying way to proceed and such a conformational experiment is quite proper.

I do not think that such a test is really required anymore just to settle the question as to whether rainmaking makes rain. Some commercial operators now use radar for evaluation. They can see the rain as it develops. So confident are they that they use the radar to direct the efforts of their aircraft and to readjust the position of the aircraft to direct the rain where they want it. We have gone well past that point. What we need now are ways of designing the whole program so that it is self evaluating. Thus we can get more information each time we work.

Much less information is necessary for decision making than is necessary for a scientific proof. Indeed most scientists know a good deal more than they can say because they simply cannot dedicate the time to proving each point to the satisfaction of pedants, even though they should. It is possible to reach correct conclusions without the formalism. Indeed, intelligence analysts frequently reach the right conclusions with wrong information. I have a friend who has developed a mathematical proof that it is possible to reach a correct answer with the wrong information by using correct information processing. I don't understand his arguments, but I approve of them. If one is careful about his cloud physics and knows something of communications theory he can tell if his work is successful without an elaborate formalistic trial. I believe that Mr. Elliott will discuss this matter in his talk. I assure you that we would both like to be able to carry out such a formal trial.

Let us assume for the moment that we have decided to go ahead and conduct such a formalistic test. We are required to set it up in such a way that we violate none of the conditions set forth earlier. This doesn't sound hard but let's think about it for a minute. Can we do it? Will we ask the right questions? Will we recognize the answers?

I can, if given a little experience in an area, cause a statistical test to succeed or fail by the way I conduct the seeding. Let us assume a simple case. We will pick cumulus clouds in a given area and seed some and not others. We will pick them at random. I won't know which are really seeded and which aren't but I can tell within 3 to 5 minutes if a cloud has been nucleated. If I notice that a seeded cloud is doing what it should, I can suppress growth of the control clouds at a considerable distance by making the seeded cloud grow. If the control cloud is close enough I can make it grow by seeding the target cloud. If we are using discrimination based on area, I can, by carefully seeding, move the mass of a cloud upwind or cross wind for a considerable distance. If we seed alternate bands in a California storm, I can affect the results of the precipitation of the next band yet to come because the wind field changes during seeding. In fact because the bands travel somewhat slower than the wind, I can affect the downwind output of clouds in the same way and thus affect the behavior of a band that already passed.

Thus, it would have to be a very carefully thought out experiment if it were not to be affected by what a skilled cloud seeder could do, while following instructions. If one wished to fail he could seed a cloud too much too soon. Indeed, this is the most common mistake made by cloud seeders.

Even with all honest intent, the above things can be made to happen and the person doing it may not know he is doing so. It is virtually impossible to design around these contingencies because one cannot foresee them all. During Project Hotshot in Oklahoma, Dr. Booker used a nuclei detector and tracked plumes of nuclei from the evaporating showers around the base of seeded clouds. He found that they went crosswind on occasion and that clouds some distance to the side were being seeded by the leftover nuclei from the evaporating showers.

Let's try what Bowen did in Eastern Australia. (See Figure 2.) He seeded the same area one year and not another, hoping to show that seeded years rained more than nonseeded years. This did not work out very well because although the seeded years did produce more rain, the nonseeded years did better than they had before the experiment started. Gradually, the whole system sort of asymptotically approached a rather considerable increase in rainfall, but the seeded years did not look much better than the subsequent nonseeded ones. This is a real problem. Either the weather changed, possibly on a two-

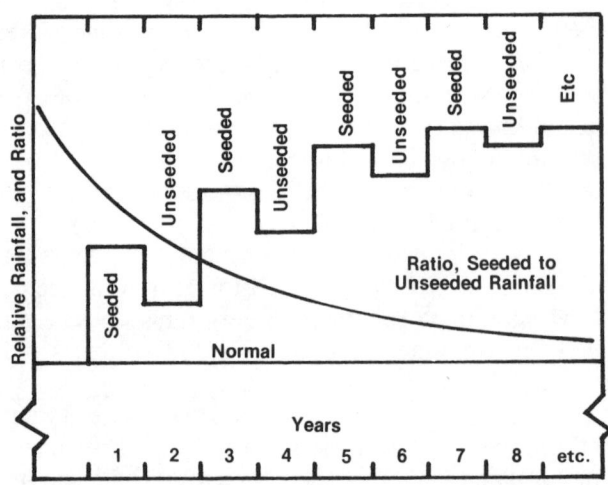

Figure 2 Conditions observed by E. G. Bowen in Australia. Ordinate is not to scale. Rainfall increased by 25% during the first seeded year. The next unseed year was well above average, as were all the succeeding years. The ratio decreased however because the overall rainfall increased asymptotically to what it was on a good seeded year. Bowen has suggested that this may be due to a modification of the land surface, and good reason exists to think so. On the other hand systematic change in the weather could have taken place.

year cycle, or more likely, Taffy had caused a change in the soil and vegetative conditions so that the local rainfall was beneficiated, as he suggested. On a smaller scale, we observed while working in the Philippines, that when an Island was thoroughly seeded on one day, it was not necessary to return for several days because rainfall continued to be high for at least four days. We were thus able to wet an island down really well and then work over others for the next few days before returning to that island. "Persistence effect" is possibly an example of how land management affects weather. Yet we might well miss this point—one considerably more important than the one we set out to test.

It is possible to seed clouds and detect the seeded areas on satellites. We were able to do this when we were seeding in Baja California Sur for the government of that state. The clouds we seeded, and no others, would grow to enormous sizes and be clearly visible on satellite photographs. It was possible to see changes in the wind field take place hundreds of miles away, caused by the intense convergence developed at low levels by the seeding. Thus a control area might have to be quite a ways away to be independent and moreover it might not even be in a reasonably similar meteorological regime.

The effects of wind shear are among the more surprising factors in cloud seeding. (See Figure 3.) Clouds in a temperature profile such that they will grow nimbly on a warm day will not necessarily grow at all if the wind shear exceeds a certain amount. It is much harder to make a cloud grow in a sheared wind field, if the shear is monotonic with altitude, than if it is if the shear is very low. On the other hand, if the shear reverses with altitude it is quite possible that clouds will become self feeding and grow to large sizes, remaining sta-

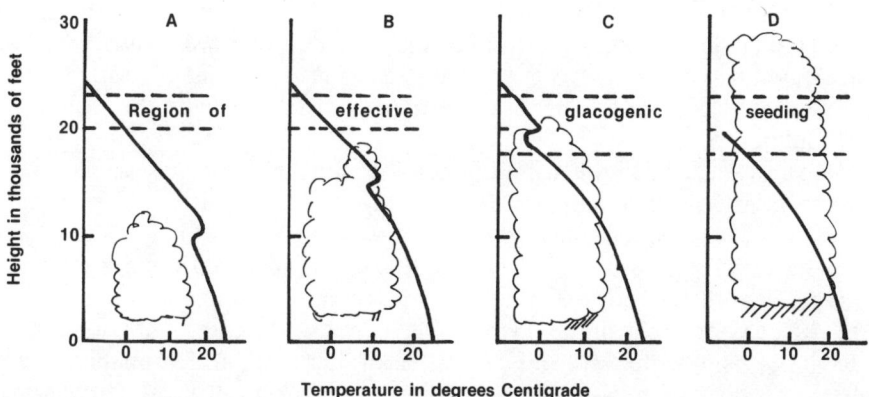

Figure 3 Temperature usually decreases, or "lapses" with height. If the temperature increases with height it is called an "inversion." Clouds will grow from the level of condensation up to a temperature inversion, but they rarely penetrate an inversion. In A, a low inversion precludes cloud growth to the "seeding window," glaciogenic seeding is useless. In B, an inversion just below the 0° level precludes seeding unless an especially vigorous tower penetrates the inversion as sometime happens. Then careful seeding will work. In C, the inversion limits cloud growth to just above the 0° level. Seeding will release enough heat to push the cloud through an inversion of perhaps 2-3 degrees. Explosive growth often occurs. Seeding works phenomenally well. In D, cloud growth in unlimited by an inversion. Seeding helps but not as much as in C.

tionary for protracted periods and causing enormous downpours. In the analysis of results of seeding it will be well always to stratify by wind shear conditions.

The effects of slight variations in lapse rate and in humidity profile are equally important. In Southeast Asia, the Philippines and many other places in the tropics, a temperature inversion at about 16,000 feet, that is to say, the freezing level, limits the growth of clouds. Only a few turrets have enough vigor and momentum to push through the inversion, or to arrive at the freezing level. If these are seeded, the extra heat released by the seeding causes the clouds to grow vigorously, right through the inversion. It is thus easy to convert a field of relatively inactive, often nonraining clouds into very large heavily raining clouds. The success ratio of a seeder under these conditions is astounding. On the other hand, if the temperature inversion drops a few thousand feet, no amount of seeding makes any difference. If the temperature inversion rises, the clouds grow by themselves and rain. Some reach large proportions on their own. Seeding under these conditions leads to increases in rainfall if and only if the small clouds are worked over and caused to grow into large ones. Seeding the large ones may be counterproductive because although they will grow, they may produce less, but perhaps more intense rain. There's little point in beating a running horse. Moreover, because the temperature inversion at lower altitudes is usually the proximate cause of droughts in the tropics, weather modifiers are often called upon to work under these conditions. If one seeds a tower too vigorously under these conditions, it often pinches off because the increased updraft induces too much entrainment of dry air from the sides of the cloud. Therefore, paradoxically, one needs a nucleant that does not work too well or too rapidly in order to seed these clouds. For this purpose we use a slightly water soluble agent that has an activation temperature of $-0.5°C$. This agent dissolves in the larger drops but works well in the smaller ones. Thus a slow steady growth takes place. In other words, *small differences in conditions make a lot of difference in the results.* If one is aware of these differences it is easy to compensate for them and to anticipate the results with some confidence.

One must consider the condition of the atmosphere. Let us take a case where an interesting phenomenon is occurring. Assume for the moment that a cloud seeding experiment was being conducted in which total randomization was being utilized. Assume also that a stable sort of cloud regime existed so that rapid growth was not observed. Small increases in rainfall of the order of 15 to 20 percent are common from such clouds when they are intelligently seeded. Let us now put into the clouds, according to some reasonable scheme, a certain number of nuclei. Will we observe an increase in rainfall? Maybe. We really ought to consider now the possibility that some industrial area to the windward of the seeded area, far enough upstream to have the effluvia mix well with the air, is emitting a great many ice nuclei. An example could be a metal refinery, a blast furnace or a test stand for large rockets. Any number of things can produce ice nuclei. (See Figure 4.) Indeed someone once measured the ice forming nuclei around rodeo grounds and found them to be abundant. The input function to the experiment is not the one envisioned in the experiment. A variable has been added to the nucleation function. Thus, the days on which

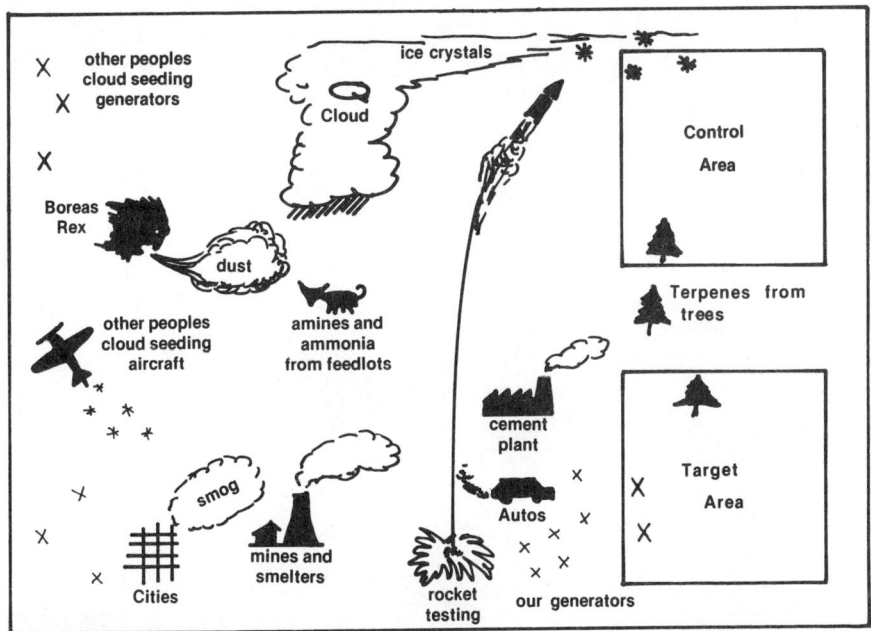

Figure 4 Seeding situation (not to scale). The contribution to the glaciogenic nuclei count made by the generators may be trivial compared to natural nuclei present in the air, to nuclei derived from industrial or urban sources and even to those from other people's seeding efforts. The effects of adding more nuclei are not linear and may even be counterproductive. A proper experiment must include a nuclei budget for freezing and for condensation nuclei.

the burners are run really have not many more nuclei or maybe less than days on which they weren't. How will the results turn out? What if a weekly function were superimposed on the random pattern? I am afraid that this has happened to one of the more important controlled experiments in the Colorado Rockies.

To be sure, it is really necessary to measure the number of nuclei in the air entering the seeded area and to be able to assess the impact that your nuclei will have upon air already containing perhaps quite enough. The volume density of condensation nuclei is also very important, in that it affects the drop size in the cloud.

Let us consider a geographical condition sketched in Figure 5 in which a randomized experiment is being carried out. A long narrow mountain range parallel to the sea is being seeded. The range is perhaps 100 miles long but only 20 to 30 miles wide. The generators are placed in the mountains and along the shore. The wind blows against the mountains from the sea, not quite at right angles, but arrive in a more or less westerly direction. An area of size equal to the target area is selected to the north and another to the south. Thus with one target area, two control areas, the seeding plume of lesser width than the target area and all other factors in hand, the experiment should work. Did it? Probably not. The following might happen.

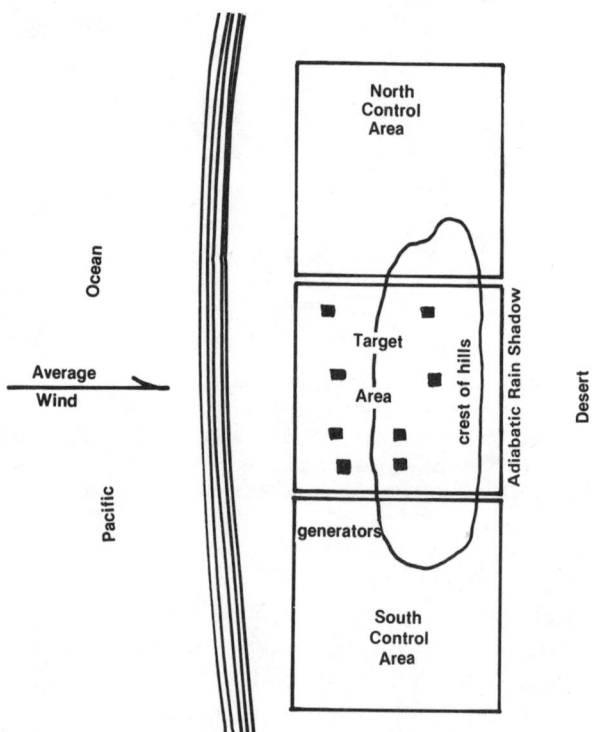

Figure 5 A seeding experiment with a double control area. A ramplike mountain range with a very steep easterly slope occupies part of the target area. Average wind is westerly but actual wind may vary widely in direction. Transit time of upper wind across mountain barrier is too short to permit growth and fall of snow crystals or precipitation. Precipitation is evaporated on far side of mountains. Lower level wind can go either to north or south, depending on the direction of the incoming air flow. Both control areas are inadvertently seeded. Target area is also seeded, but not as well.

The wind blew in from the coast, and at higher altitudes did indeed blow across the mountainous barrier. The clouds were seeded with a part of the silver iodide plume, the droplets turned to ice and continued to grow, just as a good frozen drop should. It never landed on the mountains as rain because the mountain chain was too narrow and the droplets eventually evaporated when carried into the adiabatically heated region on the lee side of the mountain chain. Some drops fell and a little increase occurred. But this is only part of it.

The wind did not always blow directly across the mountains. It cut across them at a slight angle and on some occasions caused a wind drift to the north at lower levels, and on other occasions a wind drift to the south at lower levels. At the lower altitudes, the wind could spend a long time moving to the north or to the south and so the portion of the silver iodide plume thus carried by the lower wind was moved eventually upward into clouds to either the north or south and the material reacted just as it should have. Rain fell in one control area or the other depending upon a slight difference in the wind. Thus the average annual rainfall of the entire area was increased. That of the target area only a little bit; that of the control areas a good bit.

You can see what happened. It was clearly proved by a valid statistical test that cloud seeding does not work. Whereas in reality, it worked very well. We can't really tell because this test violated, in a subtle way, A3, 4, 5, C2.

It is easy to develop testing into a kind of bear baiting contest. This has been done. It is senseless and demeaning to continue thus. One must assume some level of integrity on the part of the tester and of the operator. The whole affair, within the framework of what is known about cloud physics, the weather and the peculiarities of the region involved, must be planned so that the hidden pitfalls do not remain hidden. The statistical test should be backed up with physical tests conducted throughout the operations so that it is known if the goals are being attained or not. The tester, who may or may not be the same person as the operator, must work with the operator. If a test of such rigor can be set up it would be nice. As it is, most have not met many of the criteria for assuring successful seeding. Some have failed because they violated certain requirements of the statistical test.

I feel that it would be a good thing to take some of the theorists and most of the statisticians up for an airplane ride during a seeding expedition. The results are clear and unmistakable. There should be no trouble establishing that seeding works because the effects are so impressive. The times that they do not work are impressive too because one can often see why. Thus, it is possible to say in advance whether something is testable or not. The frontispiece of this volume is an example. This picture was taken by Dr. Shelden D. Eliott, Jr. about 7 minutes after he had seeded the left half of the cloud with about 0.1 gram of pyrotechnically produced AgI. The left half is snowing, the right half not. This is a fair statistical test.

Sometimes one has surprises. I recall seeding a cloud over the Sierra one afternoon. The cloud was a narrow vertical cloud with base at about 8,000 feet and the top at 12,000 feet. We seeded it and found ourselves in a 3,000 foot per minute updraft. The cloud base rose rapidly, the top rose, we got out of it at about 20,000 feet to find a region of considerable turbulence with a violent wind shear. The cloud base continued to rise and the whole thing spread out as a sheet at about 20,000 feet and blew off to the east. I am sure that the seeding made it grow. I am equally sure that it was destroyed by the seeding because it rose high enough to be carried off into the wind shear.

On another occasion I went on a flight with a pilot who worked for David Merril, an old time cloud seeder from Taft, California. He was using an electric arc device that, although scarcely to be described as very scientific, at least produced a good grade of silver iodide. We had seeded a number of clouds. We flew through them and seeded the updrafts. All these clouds grew. He then pointed out a tall, thick cloud about a mile in diameter and said, "Watch me make this one go away." He then flew around the periphery of the cloud, with the burner going, at about the -6 °C level. The cloud immediately disappeared. He explained that he did this every once in a while for fun. I believed him. He had called his shots in advance. What probably happened was that he initiated freezing in a down draft around the cloud and the release of heat had destroyed the down draft, or at least had slowed it down. This discouraged the updraft in the center of the cloud.

Surprises are not new to anyone who goes outdoors. They might well be to one who studies a subject without being personally involved. My real point is that a familiarity with the subject matter, at close hand, permits an understanding and hence an opportunity to learn. That is what is really important and what is too often lost. We must so arrange our experiments that we can learn from them, not just decide yes or no to a question that is in reality a whole series of subquestions and whose answer is maybe yes or maybe no or maybe that it all depends.

It is important to ask why and to ask how you know. Science really grows best when a well conducted experiment fails and someone can explain why. Without the questioning we do not really learn. Some primitive tribes have never come to appreciate the close relationship between sexual activity and childbirth. I imagine that, without some inquiry into such matters, an otherwise competent statistician whose sex education had been woefully neglected could easily arrive at the conclusion that the relationship had not been proved. After all nine months elapse between the events but even this is hard to tell because many times nothing whatsoever happens nine months later. Nonetheless, most people believe that such a relationship exists because a long link of observational evidence has been established. We perhaps can do the same with weather modificiation.

Francis Galton once said, "It is always well to retain a clear geometric view of the fact when we are dealing with statistical problems, which abound with dangerous pitfalls, easily overlooked by the unwary while they are cantering gaily along upon their arithmetic." He was speaking of the geometry of the statistical arguments—he might equally have spoken on keeping the physical facts clearly in mind. If I personally have a quarrel with anyone, it is not with members of a particular discipline—and certainly not with statisticians in particular — it is with those who do not consider the physical situation in the design and execution of weather modification experiments.

In closing, I would beseech your help in designing tests so that they were not mere exercises in arithmetic. I would beseech you to learn a bit about the subject you are treating, perhaps then you will be able to help us learn. I would also beseech you to remember that if one seeding experiment produces indeterminate results it does not mean that weather modification is not effective. It just means that that particular technique did not work under those conditions. If we can find out why we will all learn. If we are just scornful, pompously negating the effectiveness of weather modification in general, we may temporarily gain the respect of our academic peers but we will wind up less knowledgeable about the world than we could be, and perhaps be forced to conclude along with deCampoamor that:

> *Y es que en el mundo traidor*
> *Nada hay, que, es verdad ni mentira*
> *Todo es segun el color*
> *Del cristal con que se mira.*

BIBLIOGRAPHY

Aufmkampe, H. J. and Weikmann, H. K. (1951), The effectiveness of natural and artificial aerosols as freezing nuclei, *J. Met.*, 12 68-73.

Chu, F. M. (1976), *Jap. Seism. Soc.*, 11.

Cuilong, B. H. (1947), Sublimation in a Wilson cloud chamber, *Roy. Soc. London Proc., Ser A*, 190, 137-143.

deCampoamor, Ramon, *Las Dos Linternas*, written between 1817 and 1901.

Hocking, L. M. (1959), The Collision efficiency of small drops, *Roy. Meteorol. Soc. Quart.* J, 85, 44-50.

Langmuir, I. (1948), The production of rain by a chain reaction in cumulus clouds at temperature above freezing, *J. Meteorol*, 5, 175-192.

Mathews, L. A., Reed, D.W., Amand, P. St. and Stirton, R.J. (1972), Rate of solution of ice nuclei in water drops and its effect on nucleation, *J. Appl. Meteorol.*, 11, 813-817.

Odencrantz, F. Kirk (1969), Freezing of water droplets, nucleation efficiency at temperatures above $-5°C$, *J. Appl. Meteorol.*, 8, 305-346.

Raleigh, C. B. (1976), Prediction of the Haichung earthquake report by Liaoning earthquake study deleg. U.S., U.S. Department Interior Geology Survey, Conference 1, Abnormal Animal Behaviors Prior to Earthquakes, Menlo Park.

Schaefer, V. S. (1949), The formation of ice crystals in the laboratory and the atmosphere, *Chem Rev*, 44, 291-320.

Serra, Louis (1958), L'Operation Tignes (1954-1958), L'Electricite de France P. 76.

St. Amand, P., Mathews, L.A., Reed, D.W., Burkardt, L.A. and Finnegan, W.G. (1971), Effects of Solubility on AgI Nucleation Effectiveness, *J. Weather Mod.*, 3, 106-110.

Suits, C. Guy and Way, Harold E. (1960), *The Collected Works of Irving Langmuir*, Pergamon Press, N.Y. (1960).

Turnbull, David and Vonnegut, Bernard (1952), Nucleation Catalysis, *Ind. Eng. Chem.*, 44, 1292-1298.

Vonnegut, B. (1949), Nucleation of supercooled water clouds by silver iodide smokes, *Chem. Rev.*, 44, 272-289.

Some Approaches to Statistical Analysis of a Weather Modification Experiment

Ralph A. Bradley, Sushil S. Srivastava, and Adolf Lanzdorf*

Department of Statistics
The Florida State University
Tallahassee, Florida

Abstract

Data from a weather modification experiment are examined and a number of statistical analyses reported. The validity of earlier inferences is studied as are the utilities of various statistical methods. The experiment is described. The original analysis of North American Weather Consultants, who conducted the experiment, is reviewed. Data summarization is reported. A major approach to analysis is through the use of cloud-physics covariates in regression analyses. Finally, a multivariate analysis is discussed. It appears that the covariates may have been affected by treatment (cloud seeding) and that their use is invalid, not only reducing error variances but removing treatment effect. Some recommendations for improved design of similar future experiments are given in a concluding section, including preliminary trial use of blocking by storms.

1 INTRODUCTION

Phase I of the Santa Barbara Convective Seeding Test Program was conducted by North American Weather Consultants (NAWC) from 1967 through 1971 for the Research Department, Naval Weapons Center, China Lake, California. A concurrent study of the large-scale effects of cloud seeding was undertaken for the Bureau of Reclamation, U.S. Department of the Interior, Denver, Colorado by Aerometric Research Inc., the research affiliate of NAWC. Research on the design and analysis of weather modification experiments at the Florida State University is sponsored by the Office of Naval Research, Department of the Navy, Washington, D.C. Data from the Phase I Santa Barbara experiment, provided through the courtesy of NAWC, have been used for trial analyses. Some approaches to statistical analysis of the Phase I data are reported in this article.

---

*Adolf Lanzdorf is currently affiliated with Strategic Planning Associates, Washington, D.C.

A brief summary of the Phase I experiment follows. More detail is given in technical reports by Elliott and Thompson (1968a, 1968b, 1969, 1972) and in publications by Elliott, St. Amand, and Thompson (1971) and Brown, Elliott, Thompson, St. Amand and Elliott (1974). Two final reports were issued, one for the Naval Weapons Center and one for the Bureau of Reclamation, by Thompson, Brown and Elliott (1975) and Brown, Thompson and Elliott (1975) respectively. Both final reports include Phase II experiment results, 1971 through 1974. In this article, attention is on Phase I data because of experimental design changes and the introduction of aerial seeding in Phase II. Data collected for the Bureau of Reclamation study is used unless otherwise specified because of its augmented network of raingages.

Winter storms in the Santa Barbara region have identifiable convective cells grouped into bands, usually taking from one-half to one and one-half hours to pass over a point. Several convective bands may occur in a storm or it may be a single frontal band. The convective band was used as the experimental unit in the Phase I experiment. Criteria for the "seedability" of a convective band were established; in operation, they reduced to a wind flow requirement such that the possible effects of seeding would fall mainly in a target area and the expectation of substantial precipitation. Band detection was either by radar, confirmed through precipitation at a telemetered raingage in a control area to the west of a single ground seeding site, or through precipitation at two such telemetered raingages. A predetermined randomized decision to seed or not seed an experimental unit, a seedable convective band, was applied. Figure 1, taken from Elliott, St. Amand and Thompson (1971), depicts the general experimental set up; not all raingages used are shown nor were all raingages always in operation. Seeding in Phase I was ground based from a mountain crest at 1065 m. above sea level indicated in Figure 1. High output, silver-iodide, pyrotechnic devices, ignited at 15 minute intervals from the beginning of band passage for seeded bands, were used.

Band precipitation data were obtained for all raingages in control and target areas operable for a band. (The number of raingages was increased from time to time during the Phase I experimentation.) The procedure included:

(i) tracking of the precipitation band pattern across the gage network on the basis of plots of available precipitation and radar information,
(ii) determination of the time of band passage (and hence time of band duration) at each raingage, and
(iii) calculation of total precipitation from the raingage record attributable to the band.

To avoid subconscious bias, the meteorological analyst determining raingage band passage times and precipitations was uninformed as to which bands were seeded. Considerable skill was required from the analyst. A major source of variation may arise from these determinations, a disadvantage in the use of convective bands as experimental units, perhaps offset by the resulting increase in the number of experimental units available in a season.

Air mass characteristics of each band were determined from radiosonde observations at Santa Barbara Airport and Vandenburg Air Force Base (VBG in Figure 1). An attempt was always made to obtain a sounding as a band passed

Some Approaches to Statistical Analysis

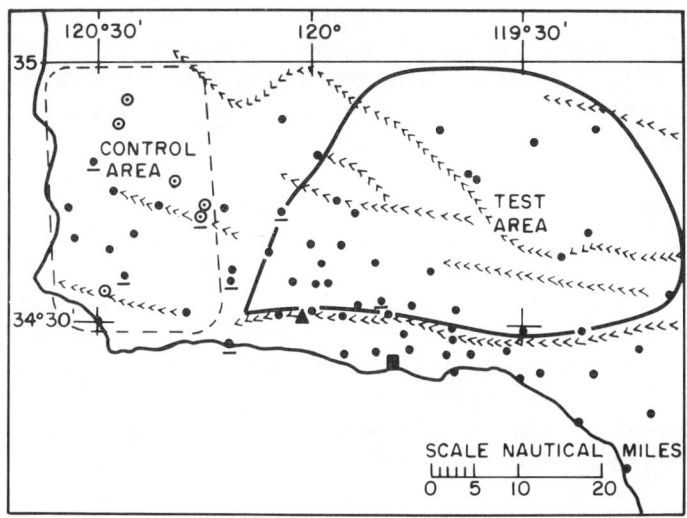

Figure 1 Santa Barbara pyrotechnic seeding and control test areas. Raingage sites are designated by solid or open circles, telemetered gages underlined. The seeding and radar site is indicated by a solid triangle. Source: Elliott, St. Amand, and Thompson (1971). Reproduced, with permission, from the *Journal of Applied Meteorology*.

over the airport. Gleeson (1977) summarized the meteorological data with a view to their use as covariates. The use of covariate analysis for the reduction of experimental error and improved sensitivity in evaluations of the effects of seeding seemed a desirable approach to evaluation of weather modification experiments. Gleeson provided data for each band (except Band 73) on the following variables:

$$
\begin{aligned}
&X_1: \text{ Mixing Ratio} \\
&X_2: \text{ 700 mb Wind Speed} \\
&X_3: \text{ 700 mb Wind Direction} \\
&X_4: \text{ Mean Wind Speed} \\
&X_5: \text{ Direction, Avg. Vector Wind} \\
&X_6: \text{ 500 mb Temperature} \\
&X_7: \text{ Stability Class} \\
&X_8: \text{ Showalter Index} \\
&X_9: \text{ Stability Wind Seed} \\
&X_{10}: \text{ Direction, Stability Wind} \\
&X_{11}: \text{ Instability Transport} \\
&X_{12}: \text{ Band Passage Time (Seeding Site)}
\end{aligned}
\tag{1.1}
$$

More detailed descriptions of these variables are given by Gleeson.

The data array for the Phase I experimentation may be viewed as a data matrix with N rows or bands, the first N_1 rows for unseeded bands and the second N_2 rows for seeded bands, $N = 107$, $N_1 = 51$, $N_2 = 56$, and with columns containing precipitation responses at individual raingages, possibly grouped by locations, and values of the concomitant variables, X_1 to X_{12}. The data are not without problems. Raingage precipitation responses are correlated (correlation approximately 0.6) as would be expected. There are missing data for many gages. There may be problems also in consideration of rows as independent observation vectors. Elliott and Thompson (1968b) consider the persistence effects of seeding and conclude: "some seeding precipitation enhancement may have occurred in nonseeded bands which follow on seeded bands." NAWC analyses, Elliott and Thompson (1969), suggest the possibility of an up-wind effect west of the seeding site attributable not to westward seeding contamination but to seeding-caused blocking of the air-mass flow leading to up-wind convection development. Bradley, Srivastava, and Lanzdorf (1977a) provide precipitation summarization data used below. These data, together with those of Gleeson, are available to readers interested in investigating other approaches to the analysis of this weather modification experiment.

Primary NAWC analyses are reviewed in the next section. This is followed by a short discussion of the authors' efforts to summarize the precipitation data. The use of the available data in regression-covariance analyses is reported, followed by a preliminary multivariate analysis. The article concludes with some remarks on the design of similar, future weather modification experiments.

2 NAWC DATA ANALYSIS

The main NAWC approach to data analysis was on a raingage station-by-station basis.

Let $y_{i\alpha}$ denote precipitation at station i from band α, $\alpha = 1, \ldots, N$. Let $\gamma_\alpha(i) = 1$ or 0 as station i was operative or not operative for band α and let $\delta_\alpha(i) = 1$ or 0 as band α was seeded or not seeded. Then $\Sigma_\alpha \gamma_\alpha(i) = N(i)$ and $\Sigma_\alpha \delta_\alpha(i) \gamma_\alpha(i) = N_s(i)$, respectively the number of observations and the number of seeded bands recorded at station i. The number of unseeded bands at station i is $N_{ns}(i) = N(i) - N_s(i)$. Then

$$\bar{T}_s(i) = \sum_\alpha \delta_\alpha(i) \gamma_\alpha(i) y_{i\alpha}/N_s(i) \qquad (2.1)$$

and

$$\bar{T}_{ns}(i) = \sum_\alpha [1 - \delta_\alpha(i)] \gamma_\alpha(i) y_{i\alpha}/N_{ns}(i) \qquad (2.2)$$

are precipitation averages at station i for seeded and nonseeded bands respectively. Six control area detection stations were used, stations circled in Figure 1. If k indexes these control stations, define

$$\bar{C}_s = \sum_k \bar{T}_s(k)/6 \qquad (2.3)$$

and

$$\bar{C}_{ns} = \sum_k \bar{T}_{ns}(k)/6. \qquad (2.4)$$

The descriptive statistic used by NAWC for station i, a double ratio, is

$$DR(i) = [\overline{T}_s(i)/\overline{C}_s]/[\overline{T}_{ns}(i)/\overline{C}_{ns}]. \qquad (2.5)$$

Use of the double ratio was compared with use of the single ratio,

$$SR(i) = \overline{T}_s(i)/\overline{T}_{ns}(i). \qquad (2.6)$$

It was found that much the same results were obtained for the two statistics. The intent in use of (2.5) was to standardize the comparison of seeded and nonseeded responses through divisions by control area precipitations, assumed to be unaffected by seeding. If the control area detection stations had missing observations, apparently $\overline{C}_s$ and $\overline{C}_{ns}$ in (2.3) and (2.4) were evaluated from the available observations.

Figure 2, based on the Naval Weapons Center study, shows contours for the double ratio of (2.5) for the Phase I experiment. Similar figures are given by Elliott and Thompson (1972) for the single ratios of (2.6) and for subdivisions of the data by years, stability classes, and 500 mb temperatures. The locations of regions of possible precipitation enhancement are fairly stable in all such figures and they tend to be regions with low average precipitations for both seeded and unseeded bands.

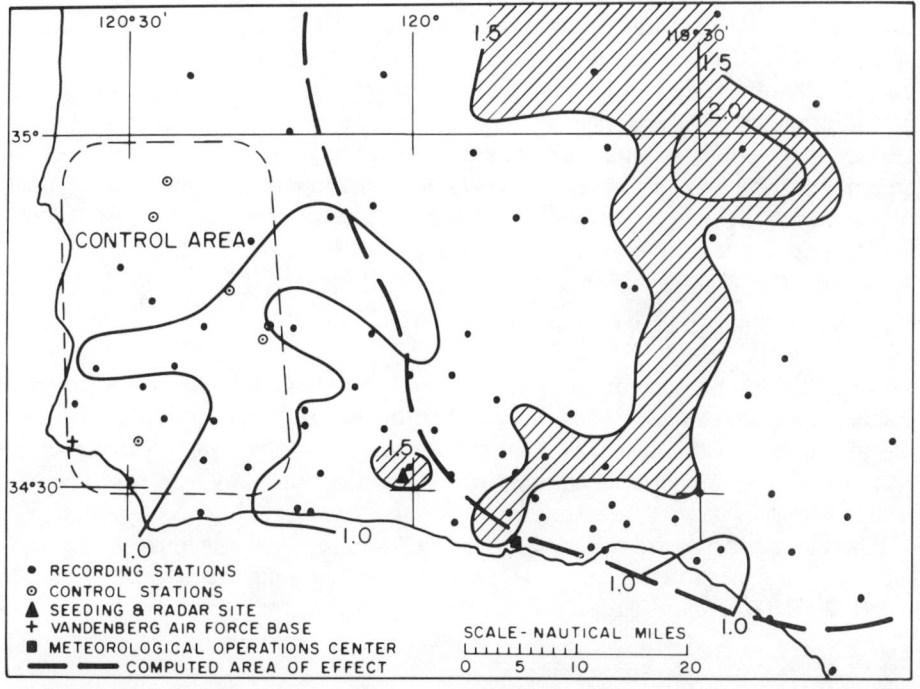

Figure 2 Composite double ratios of precipitation for 1967-71, all bands, ground-seeded, 56 seeded and 51 not seeded. Source: Elliott and Thompson (1972).

NAWC states in their various reports that the Wilcoxon-Mann-Whitney, two-sample, rank test was used to assess the significance of double and single ratios for each raingage station. The methods of application of the test are not clear in the reports but we give our understandings. For the single ratio, $y_{i\alpha}$ was used; the precipitation measurements themselves were grouped into two samples, seeded and unseeded, and the rank test applied. For the double ratio, $y_{i\alpha}/\overline{C}_\alpha$ was calculated for each band α at stations i, $\overline{C}_\alpha$ being the average of the six (or available) control-area, detection stations for band α, and these indices were grouped into two samples as before and the test applied for station i.

Similar analyses were completed by NAWC using band duration times instead of precipitations as the responses and very similar results were obtained. See Figure 5.12 of Elliott and Thompson (1972). We concentrate on precipitation analyses in this article, but the possibility that seeding may affect duration time is discussed below.

NAWC was aware that their analyses were open to possible criticisms. Possible persistence effects of seeding would seem to decrease the apparent effects of seeding and raise questions about the use of convective bands as indpendent experimental units. Station-by-station tests of significance are not independent. The responses $y_{i\alpha}$ for the various stations i have correlations of aproximately 0.6 as noted above. The indices $y_{i\alpha}/\overline{C}_\alpha$ have additional dependencies because the denominators are the same for all stations for band α. Concerned with these dependencies, NAWC conducted a limited Monte Carlo study, reported by Elliott and Brown (1971) in their Table 1, to give additional credence to their conclusions. They state: "At the 0.05 significance level for all bands, 29 stations in the original test sample were found to show a positive difference between seeded and not-seeded cases; and three Monte Carlo runs (out of 50) were found to have as high or higher counts of stations with a positive difference at this significance level." The use of ratios to measure precipitation enhancement is open to question depending on project objectives. If large ratios occur in areas of relatively low precipitation, somewhat sparsely represented by raingages, the effect on total or average precipitation for a larger defined target area may be small and the result of little economic value.

3 DATA SUMMARIZATION

A more direct approach to the analysis of a weather modification experiment is to consider summary measures of precipitation for each experimental unit over designated response areas. The arithmetic mean of the raingage measurements over a response area for each unit would be the summary measure typically used.

Bradley, Srivastava, and Lanzdorf (1977a,b) defined response areas as in Table I. The locations of these areas may be identified through reference to Figure 2. The first five areas will be referenced as target areas and the last as

Table I Definitions of response areas.

Response Area	Ranges in Degrees Latitude	Longitude	Number of Stations
(i)	34.0-35.25	118.0-120.02	107
(ii)	34.4-35.0	119.51-120.02	26
(iii)	34.0-35.0	118.0-119.51	72
(iv)	Areas (ii) + (iii)		98
(v)	All Stations in the Naval Weapons Center Reports East of Seeding Site, long. 120.02		61
Control[a]	34.4-35.25	120.02-120.60	34

[a]The control area for the Naval Weapons Center study consists of all 39 stations west of the seeding site.

the *control area*. The number of raingage stations and the data used for Target Areas (i)-(iv) are those of the Bureau of Reclamation study and those for Target Area (v) are those of the Naval Weapons Center study with minor modifications noted in the two cited references. Note that these target areas cover the test area of Figure 1, but that some raingages existed outside of these response areas, some of them in arid regions.

Precipitation averages in inches are exhibited in Table II for the various response areas. They were computed as simple averages of the individual convective band averages of available raingage measurements for the band in the designated response area. The numbers of raingages available increased

Table II Precipitation means in inches.

Response Areas	(i)	(ii)	(iii)	(iv)	(v)	Control
Seeded Bands	0.257	0.329	0.249	0.271	0.267	0.234
Unseeded Bands	0.178	0.229	0.172	0.187	0.190	0.203

somewhat with the seasons and not all raingages were operable for all convective bands. Table II is intended only to indicate the nature of responses. It reinforces impressions given by Figure 2 with its double ratios. The control area mean for seeded bands is higher than that for unseeded bands, as are target area means, suggesting either that seeding had some effect in the control area or misfortune in the randomized choices of bands to be seeded.

With the intent of improvement of data summarization, Bradley, Srivastava, and Lanzdorf (1977a,b) summarized the precipitation data through the use of response surfaces for the control area and target area (i) separately. The basic independent variables were the coordinates of latitude and longitude for the raingage stations with individual, raingage precipitation measurements as the dependent variable observations. Separate response surfaces were found for each convective band. It was found necessary to use general cubic response models to represent responses adequately. Precipitation volumes and their variances were calculated, the volumes obtained through integrations of the

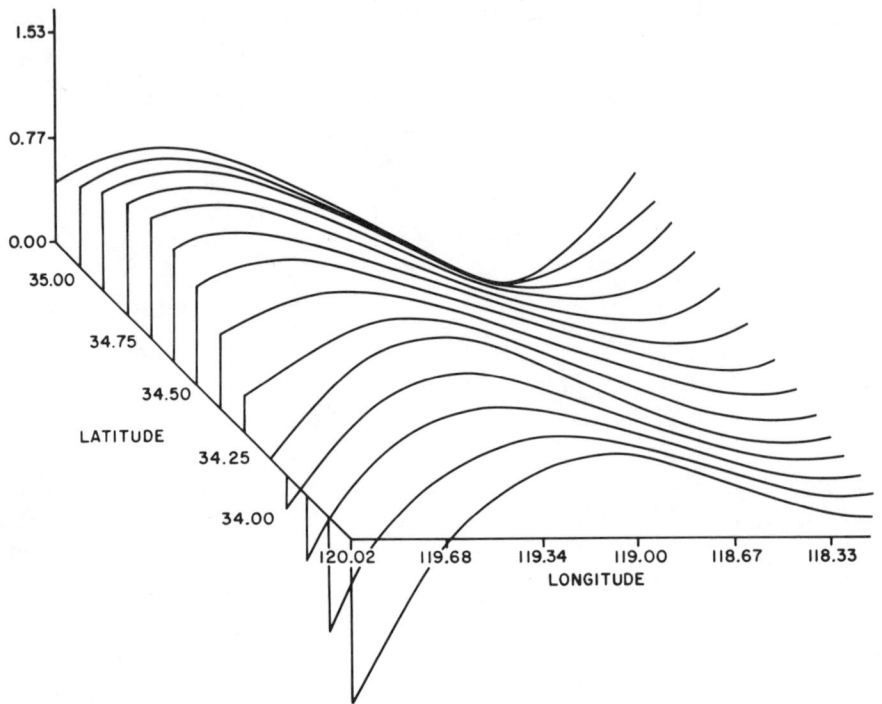

Figure 3 Graph of Cubic Response Surface: Band 96 (Seeded), Target Area (i). Vertical axis is 2.3 times precipitation in inches. Source: Bradley, Srivastava, and Lanzdorf (1977b). Reproduced, with permission, from American Meteorological Society: Fifth Conference on Probability and Statistics.

surfaces over the designated target areas or control area. Figure 3 is typical of results obtained; the region where the surface is negative is offshore.

The response surface approach was successful as a method of data summarization. It was not successful in improvement of data summarization in comparison with use of the raingage means over stations within response areas for a convective band. Some 70% of the inherent variation in responses among raingages within a band and response area was explained by the independent variables, the percentages varying considerably from band to band. Residuals from fitted surfaces exhibited only limited spatial trends when the cubic surfaces were used. Correlations between precipitation volumes calculated from the response surfaces and precipitation means were given by Bradley, Srivastava and Lanzdorf (1977a,b). They ranged from 0.97 to 0.99 for Target Areas (i)-(iv) and the correlation was 0.89 for the control area. The use of volumes in consideration of the effects of seeding cannot be expected to yield additional insights, although Bradley, Srivastava, and Lanzdorf (1978) did examine their use as reported in the next section.

Scott (1978) used a multivariate approach to data summarization. He found principal components among raingage responses in both Target Areas (i) and the control area with a view to summarizing responses through one or more orthogonal linear combinations of the raingage measurements of stations in response areas for each band. Thus, raingage responses were treated as variates and convective bands as experimental units. Substantial pruning of the data and some innovations were required to circumvent the serious problems of missing observations in multivariate analysis. Correlation matrices were obtained and principal components determined, along with their eigenvalues. The first three principal components were interpretable approximately as a mean response, a coastal versus inland contrast, and an east-west contrast. Percentages of variation explained by these components were respectively 71.3, 6.7 and 5.9 in Target Area (i) and 76.1, 6.7 and 4.7 in the control area. The correlations of the first component with the band mean were 0.997 for Target Area (i) and 0.985 for the control area. Scott is engaged in use of these results in examination of the effects of seeding. The first component cannot be expected to yield new insights; other components may add some new information.

4 COVARIANCE ANALYSES

Weather modification experiments are conducted necessarily in a natural environment involving much variability. The use of covariates in analyses for the reduction of experimental error appears to be the major available means to improved experimental design. It was for this purpose that Gleeson (1977) summarized information on covariates as discussed in Section 1. We report in this section on covariance analyses effected through use of multiple regression methods.

Bradley, Srivastava, and Lanzdorf (1978) reported on initial covariance analyses. (Some later analyses are reported below.) Regression models used were of the form,

$$U = \beta_0 + \sum_{i=1}^{p} \beta_i V_i + \delta Z + \epsilon \qquad (4.1)$$

where U is a precipitation response variable for a target area, V_i is the ith covariate, $Z = 1$ or 0 as the experimental unit was or was not seeded, the β's and δ are regression parameters, and ϵ is a random error. The data matrix has rows, $(U_\alpha, V_{1\alpha}, \ldots, V_{p\alpha}, Z_\alpha)$, $\alpha = 1, \ldots, N$. The regression parameters were estimated by weighted least squares through minimization of

$$\sum_{\alpha=1}^{N} w_\alpha (U_\alpha - \beta_0 - \sum_{i=1}^{p} \beta_i V_{i\alpha} - \delta Z_\alpha)^2. \qquad (4.2)$$

In the referenced report, use of the set of covariates of (1.1) and their interactions with treatment (seeding), along with X_c, a measure of control area precipitation, was explored. Both target area mean precipitation and target area precipitation volume, see Section 3, were used as U for Target Areas (i)-(iv), together with corresponding measures for X_c. No results are summarized here for precipitation volumes since they were very similar to those for mean precipitations. Pairwise unweighted correlation coefficients are shown in Table I of the Appendix to this article (Table A-I) for mean precipitation, X_c, and the covariates of (1.1) to given an indication of relationships for Target Area (i). Note that X_c, control area mean precipitation, and X_{12}, Band Passage Time, correlate most highly with target area mean precipitation; both of these covariates may be affected by seeding — we have noted a possible effect of seeding up-wind from the seeding site in the control area and it has been conjectured that the effect of seeding may be to increase rainfall through an increase in band passage time.

It was reported in the reference, after preliminary analyses, that seven of the twelve covariates of (1.1) were sufficient for experimental error reduction. They were:

$$X_2: \text{ 700 mb Wind Speed}$$
$$X_3: \text{ 700 mb Wind Direction}$$
$$X_6: \text{ 500 mb Temperature}$$
$$X_7: \text{ Stability Class} \quad (4.3)$$
$$X_8: \text{ Showalter Index}$$
$$X_{11}: \text{ Instability Transport}$$
$$X_{12}: \text{ Duration of Band Passage}$$

The selection was based on redundancy considerations and their contributions to error reduction. Final analyses were done for four models with unit weights (unweighted) and weights, $w_\alpha = n_\alpha / s_\alpha^2$, where n_α is the number of raingage observations contributing to the precipitation mean for band α and s_α^2 is the variance among those observations. The models in the form (4.1) had the covariates V_i as follows:

Model	Identification of $V_1, \ldots, V_p$
(1)	$X_2, X_3, X_6, X_7, X_8, X_{11}, X_{12}$
(2)	V's of Model (1) plus $X_2 Z, X_3 Z, X_6 Z, X_7 Z, X_8 Z, X_{11} Z, X_{12} Z$
(3)	X_c plus V's of Model (1)
(4)	X_c plus V's of Model (2)

(4.4)

Values of the coefficient of determination R^2, the square of the multiple correlation coefficient for N = 106 bands*, are given for the four models and Target Areas (i)-(v) in Table III.

Table III Coefficients of determination (R^2) for regressions with precipitation means.[a]

Target Area	Models Without Control Mean				Models with Control Mean			
	Unweighted		Weighted[b]		Unweighted		Weighted[b]	
	(1)	(2)	(1)	(2)	(3)	(4)	(3)	(4)
(i)	0.597	0.621	0.364	0.437	0.712	0.750	0.593	0.615
(ii)	0.578	0.608	0.373	0.457	0.789	0.815	0.681	0.691
(iii)	0.578	0.606	0.285	0.468	0.646	0.691	0.505	0.582
(iv)	0.604	0.629	0.371	0.442	0.712	0.751	0.589	0.610
(v)	0.575	0.593	0.344	0.426	0.778	0.805	0.603	0.627

[a] Values of R^2 in the reference for weighted regressions have been corrected.
[b] See Table A-5, Bradley, Srivastava, and Lanzdorf (1978).

---

*See Gleeson (1977); there were 107 bands but covariate data were missing for Band 73.

Bradley, Srivastava, and Lanzdorf (1978) gave estimates of the regression parameters and corresponding analysis of variance tables with sources of variation being reduction in variation due to basic covariates, additional reductions due to interactions (when included in the model) of the basic covariates with seeding, final reduction due to seeding, and residual variation. Results were disappointing. There were no apparent effects due to seeding. There was little interaction. The combined effects of the basic covariates were significant, generally at the 0.01 level of significance.

We were not satisfied with the preliminary analyses. Standard deviations were related to means as seen in Figure 4 below for Target Area (i). Values of n_α varied also. The weighted analyses gave very heavy weights to bands with low precipitation means; values of R^2 were reduced as seen in Table III and weighted means were quite different from the unweighted means of Table II, often suggesting more precipitation for unseeded bands. We report now on new analyses with the data transformed to stabilize variances.

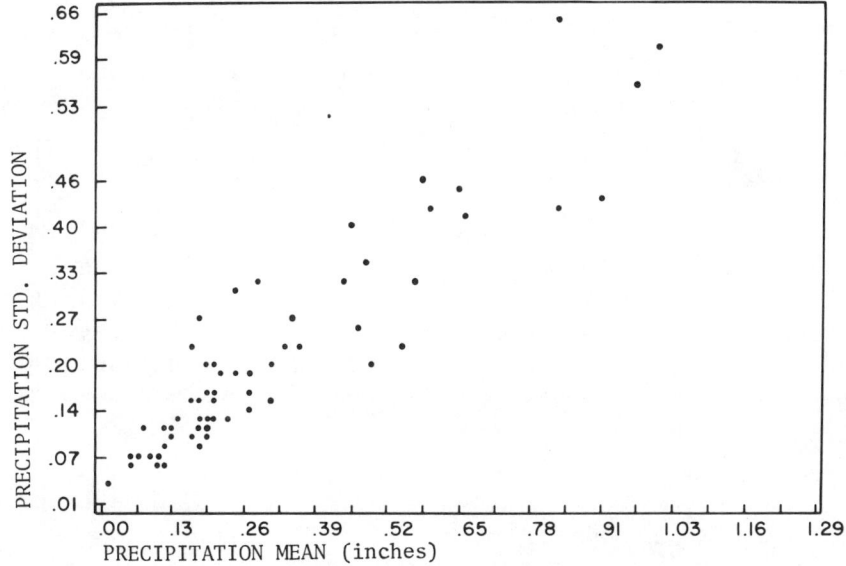

Figure 4 The relationship between mean and standard deviation, target area (i).

Analysis of the data of Figure 4 and similar data for the other target areas suggested the use of logarithmic transformations to stabilize variances. Given a raingage observation y, the transformed responses were of the form, log (1+ay). Correlations with U_2, the target area mean of the transformed responses, are shown in Table A-I for Target Area (i); they are very close to those for U_1, the target area mean precipitation. Figure 5 shows the standard deviations of the transformed responses plotted against values of U_2 for Target Area (i). It is seen that variances have been stabilized except for small values of U_2, values for convective bands that may not have been acceptable "seedable" bands.

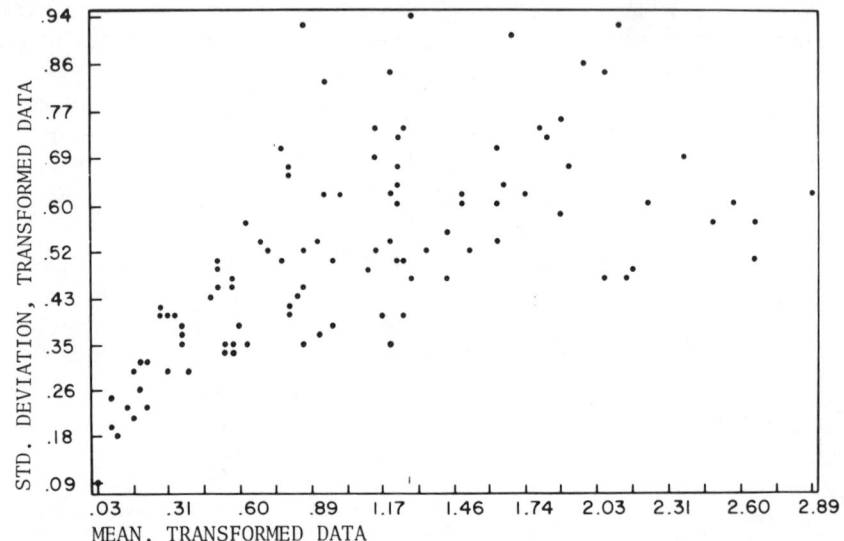

Figure 5 The relationship between mean and standard deviation, transformed data, target area (i).

Regression analyses similar to those described above were done. Models with and without X_c and X_{12} were used because it had been suggested that they may have been affected by seeding. The response variable for each model is the mean of the transformed precipitations noted above for the designated target area and band. The weights w_α in (4.1) were taken to be n_α, the number of raingages operative in the target area for band α. The models used were as follows:

Model	Identification of $V_1, \ldots, V_p$
(5)	$X_c, X_2, X_3, X_6, X_7, X_8, X_{11}, X_{12}$
(6)	Model (5) less X_c
(7)	Model (5) less X_{12}
(8)	Model (5) less X_c, X_{12}
(9)	Model (5) plus $X_2Z, X_3Z, X_6Z, X_7Z,$ $X_8Z, X_{11}Z, X_{12}Z$

Results from these newer regression analyses are summarized. Table IV shows values of R^2 that may be compared with those of Table III. In particular, values for models (5) and (6) of Table IV may be compared respectively with those for models (3) and (1)

Table IV Coefficients of determination (R^2) for regressions with means of transformed precipitations.

Target Area	Models (5)	(6)	(7)	(8)	(9)
(i)	0.720	0.616	0.659	0.293	0.748
(ii)	0.769	0.577	0.732	0.236	0.788
(iii)	0.663	0.603	0.585	0.299	0.696
(iv)	0.725	0.632	0.657	0.302	0.752
(v)	0.741	0.558	0.709	0.229	0.764

of Table III. Slightly larger values of R^2 were obtained with the transformed data. Results for model (8) show that R^2 is reduced considerably when X_c and X_{12} are omitted as covariates. Model (9) has values of R^2 similar to but slightly larger than for Model (5). Table A-II contains the essentials of analysis of variance tables for models (5)-(9) for the five target areas of Table I based on the transformed data. It is seen that use of X_{12}, duration of band passage and X_c, control area mean precipitation, to a lesser extent, as covariants reduces the apparent effect of seeding; F-ratios for seeding are largest for model (8) without inclusion of either X_c or X_{12}. Regression coefficients for model (5) are given in Table A-III for all five target areas. These values permit reconstruction of the estimated regression models and reinforce comments relative to X_c and X_{12} above. Examination of residuals about estimated regression models for the transformed data suggests that transformation improved symmetry and approximate normality of their distributions.

Gleeson (1977) saw no major differences between results on covariates for seeded and unseeded bands but he did observe that the differences exhibited some consistency. After some discussion, he wrote: "Taken separately the effect of these differences may be insignificant, but in combination they suggest that the total precipitation that might have been realized from seeded bands, had they not been seeded, would have been larger than the total amount that fell from nonseeded bands." Gleeson's concern could be explained by unfortunate randomization in the seeding decision or through a seeding effect on the covariates. We are inclined to the latter possibility. We have referred to Elliott and Thompson (1969), who raised the possibility of an up-wind effect in the control area that would affect X_c. Brown and Elliott (1972), in discussing the time duration of a seeded band, state: "There is some evidence that this increased duration is caused by a slowing down of the back edge of the band as it moves across the area of effect." This could affect X_{12}. Other covariates were measured by radiosonde at Santa Barbara Airport, well into the target area, and their values may have been affected by seeding also. Because of these concerns, we have done analyses of the transformed data omitting all covariates but retaining the weights, $w_\alpha = n_\alpha$.

Table V shows results for analyses of variance for the transformed data without use of covariates. The appropriate test should be one-sided and, in each case, the regression coefficient for seeding was positive. Assessment of $t = \sqrt{F}$ leads to a one-sided significance level of 0.05 or slightly larger for each target area. While the test are not independent, these results confirm those of the Monte Carlo assessment of the NAWC analysis of Section 2.

Table V Analysis of variance without covariates for the various target areas, transformed data.

Target Area	Source of Variation	d.f.	Mean Squares	F-Ratio
(i)	Seeding	1	110.29	2.77
	Residual	104	39.84	—
(ii)	Seeding	1	38.57	2.84
	Residual	104	13.59	—
(iii)	Seeding	1	79.49	2.52
	Residual	104	31.55	—
(iv)	Seeding	1	100.33	2.76
	Residual	104	36.30	—
(v)	Seeding	1	73.01	2.58
	Residual	104	28.32	—

We have now highlighted what may be the major design defect in the Phase I Santa Barbara experiment, perhaps a defect that could not have been anticipated. The result has been that the covariance analyses have not been helpful, and, indeed, represent a misuse of the method, one that commonly occurs. Nevertheless, we believe that covariance analysis should be a good means to improved experimental precision. In future experiments, attention should be given to choice of good covariates measured in appropriate locations, free from the effects of seeding. Perhaps measurements at Vandenberg Air Force Base, well west of the seeding site, prior to seeding, would have been suitable.

The analyses of this section are open to minor technical concerns. The persistence effect of seeding again raises questions about the independence of experimental units. Normality assumptions are not valid for individual raingage observations but should be appropriate for target area means. Variance heterogeneity is present, but should be of little concern for analyses with transformed data. Choice of weights, $w_\alpha = n_\alpha$, for analyses with transformed data is only strictly appropriate if raingage observations are independent. Independent variables in the regression models are subject to experimental errors. While these concerns are present, we do not believe that analyses should be misleading, particularly when the transformations are used.

5 A MULTIVARIATE ANALYSIS

Somewhat in the spirit of the NAWC station-by-station analysis, but without the problem of correlated univariate tests, we have performed a crude multivariate analysis without use of covariates or transformations. We report briefly here since several problems with the multivariate approach arise.

We could not treat the response from each target area raingage as a separate response since the number of raingage stations exceeds the number of experimental units and because there would be many missing observations. Target Area (i) was divided as a 3×4 grid creating 12 subareas as defined in Table VI. In that table we show simple means $\bar{y}_s$ and $\bar{y}_{ns}$ for 50 seeded and 41 unseeded bands for each subarea, together with $\bar{n}$, the number of raingages operative on average in the subarea. Values of t are given for the unvariate, two-sample Student test for subareas; one-sided significance levels vary from 0.025 to 0.149. Only 91 bands could be used because 16 bands had no operative stations in one or another of the subareas.

The multivariate approach considers the mean response per band for each subarea as one of 12 variates and the two-sample Hotelling test is applied with sample sizes of 50 and 41. The F-statistic associated with the test has the value 0.81 with 12 and 78 degrees of freedom and a two-sided significance level of 0.64, not indicative of a seeding effect, and less indicative of such an effect than any of the subarea Student tests, all of which have consistently postive values of t.

What has happened in the multivariate test? The sample dispersion matrix leads to correlations between subarea means of approximately 0.8, larger for subareas close together and smaller for subareas farther apart. In the computation of the Hotelling statistic, the quadratic form involved has a matrix in which all non-diagonal terms are negative and they are associated with cross-products of the variates that are always positive. Thus the quadratic form, when evaluated, has a value much reduced from the sum of its terms involving squared variate observations.

Table VI Subareas of target area (i), seeded and unseeded means, and values of the student statistic.

Latitudes		Longitudes			
		119.5°-120°	119°-119.5°	118.5°-119°	118-118.5°
34°-34.41°	$\bar{n}$	2.8	8.5	16.4	21.6
	$\bar{y}_{ns}$	.214	.186	.156	.168
	$\bar{y}_s$	.285	.269	.240	.266
	t	1.14	1.45	1.58	1.74
34.41°-34.82°	$\bar{n}$	18.4	7.4	4.8	4.1
	$\bar{y}_{ns}$	.228	.182	.145	.080
	$\bar{y}_s$	.364	.285	.243	.143
	t	1.97	1.80	1.90	1.74
34.82°-35.25°	$\bar{n}$	6.2	5.6	1.5	1.6
	$\bar{y}_{ns}$	.054	.038	.042	.057
	$\bar{y}_s$	.083	.065	.075	.084
	t	1.49	1.54	1.53	1.05

The high positive correlations between subarea means have reduced greatly the effectiveness of the multivariate analysis.

There are other problems with the multivariate approach. The numbers of raingage stations operative in subareas vary from band to band and hence observation vectors are not identically distributed; in particular, they do not have a common dispersion matrix. In addition, comparison of the two sample dispersion matrices for seeded and unseeded bands shows that they are not the same at a high level of significance and the multivariate Behrens-Fisher problem arises. As in all other analyses, independence of observation vectors is suspect because of possible persistence effects of seeding.

Multivariate analysis does not appear to be a likely way to improved future experiment design and analysis. Missing observations cause great difficulty and lead to exclusion of experimental units containing good information. Only the most rigorous effort to avoid missing data could obviate the difficulty. The high correlations among subarea means would require development of special methods for efficient analysis of resulting data.

6 ADDITIONAL ANALYSES, REMARKS AND RECOMMENDATIONS

After completion of the analyses reported above and development of a preliminary manuscript for this article, the reports of the Weather Modification Board (1978) and of its Statistical Task Force (1978) became available. The first emphasizes the national importance of weather modification and the need for much future research. The second addresses many of the statistical problems associated with such research, delineating between exploratory and confirmatory experimentation. Brillinger, Jones and Tukey, in the second report, emphasize the need for good covariates unaffected by seeding, blocking of experimental units, and the need for randomization analyses. We have regarded the Phase I Santa Barbara experiment as exploratory and parametric analyses as an appropriate and efficient approach to exploration of the data for new insights into improved future experimental design.

An opportunity for the blocking of convective bands by storms was not used in the design of the Phase I Santa Barbara experiment. The randomized decision on seeding was made for each experimental unit individually. Accordingly, the analyses of variance of Table V are appropriate if no covariates are used. But we may obtain some insights into the effects of blocking by storms and provide analyses in Table A-IV. There are 38 storms, some with only one experimental unit, some with several experimental units, all of which were either seeded or not seeded, and some with both seeded and not-seeded units. Storm effects were totally or partially confounded with seeding effect. The analyses of Table A-IV were done in such a way as to consider the additional effect of seeding after adjustments for storm effects. We see that the inclusion of storm effects in the model has increased values of R^2 and reduced residual or error variances (compare Table V with Table A-IV). But the apparent effect of seeding has disappeared again. In future similar experimentation, use of storms for blocking should be considered, perhaps as suggested by the Statistical Task Force, with randomization of the seeding decision within storms. In the Phase II Santa Barbara experiment, a design change led to seeding or not seeding all convective bands within a storm because of concerns for a persistence effect of seeding. We plan to do randomization analyses in confirmation of indications of Table V and A-IV, consistent with a suggestion by Gabriel elsewhere in this volume that randomizaton analyses might be reserved for the most critical comparisons.

In further exploratory analyses, we considered as additional sources of variation position of the band within a storm and a possible first-order carry-over effect of seeding from a seeded band to the following band if in the same storm. No real effects for positions or carry-over were found.

Some remarks and recommendations can be made after analysis of the Phase I Santa Barbara experiment. We are in near agreement with the conclusion of Elliott and Brown (1971): "Even when those bands not as receptive to seeding were included in the sample, the seeded to not-seeded precipitation increases were greater than 50%." The means of Table II show increases near to 50% and the analyses of Table IV suggest significance near to the 0.05 level.

Improved experimental design is needed but not easy to achieve. Use of convective bands as experimental units increases the number of available units per season but raises other problems. Some improvements are needed:

(i)   Improved detection and determination of "seedable" bands.
(ii) More uniform dispersement of raingages over regions of interest.
(iii) Improved determination and measurement of precipitations attributed to particular convective bands.
(iv) Better determination and measurement of covariates free from possible seeding effects.
(v)   Allowance for blocking by storms for further control of variation.

The quotation above supports the need for (i). Low precipitation bands in Figure 4 cause difficulty with transformations as seen in Figure 5 and these may not have been good "seedable" bands. More uniform dispersion of the raingages over the target area would be desirable, although it is understood that practical difficulties in so doing arise—many raingages used had established locations and accessability is a factor also. Better dispersion of the raingages as stated in (ii) should permit reduction of the number used. It is suspected that reading of a raingage for precipitation attributed to a given convective band is very difficult and introduces considerable experimental variability. We do not know how to effect (iii) and the difficulty is offset by the availability of more experimental units in a season when they are taken to be convective bands. The persistence effect of seeding discussed in the article is more acute with use of convective bands as experimental units.

The major design change needed in future experimentaton involves the measurement of suitable covariates, covariates not subject to possible changes due to seeding (iv). It would appear to have been better to have taken the radiosonde observations at Vandenberg Air Force Base than at Santa Barbara Airport; they would then have been taken prior to seeding and hopefully unaffected by seeding. Measurement of band passage time at Vandenberg rather than at the seeding site might have been better also. The use of a control area seemed an attractive idea but may not be feasible unless chosen to be unaffected by seeding. Further meteorological research may identify air-mass covariates more closely correlated with precipitation. Elliott, in correspondence, has suggested reasons on technical grounds for the use of nonlinear functions of the available covariates and that very careful formulation of covariates is necessary if they are to be effective. While we respect his theoretical insight, models used may be regarded as first-order approximations to more complex ones. Our use of covariates reduced experimental errors; the flaw is that they seem likely to have been affected by seeding. Blocking by storms in future experiments seems feasible as suggested in (v) and likely to be helpful, particularly if blocks of unequal sizes are used as suggested by Brillinger, Tukey and Jones.

On the statistical side, transformation of the data to stabilize variances and to improve normality seems necessary. Further investigation may lead to better transforms. Multivariate methods similar to the one used in Section 5 do not seem helpful and place too stringent requirements on the experimenter. It remains to be seen if use of principal components, as considered by Scott (1978), will be helpful. In spite of the problems encountered, we believe that covariance-regression analyses, like those of Section 4, give the most potential for improved analysis of future experiments. It will be necessary to obtain good covariates, unaffected by treatment. We do not like the array of univariate tests used in Section 2 because significance is difficult to determine. Confirmation of promising analyses by randomization tests may be desirable.

Acknowledgments

Research supported by the U.S. Office of Naval Research under Contract No. N00014-76-C-0394. Reproduction in whole or in part is permitted for any purpose of the United States Government. We are indebted to Professor Elton Scott for computations reported in Section 5 and to Professor Thomas A. Gleeson for definition and calculation of the meteorological covariates. This article was also published in a special issue of *Communications in Statistics, Part A* on statistics in weather modification.

Appendix Tables A-I through A-IV follow.

Table A-I Correlation coefficients among precipitation measures,[a] covariates, and seeding, Target Area (i).

	U_1	U_2	X_c	X_1	X_2	X_3	X_4	X_5	X_6	X_7	X_8	X_9	X_{10}	X_{11}	X_{12}
X_c	.74	.74													
X_1	.02	.05	−.01												
X_2	.38	.31	.26	.10											
X_3	−.32	−.35	−.20	.13	−.00										
X_4	.37	.36	.22	.20	.76	−.15									
X_5	−.32	−.34	−.20	.06	.07	.90	−.08								
X_6	−.12	−.14	−.01	.64	.28	.22	.26	.14							
X_7	−.10	−.11	.03	−.06	.24	.04	.04	.05	.23						
X_8	−.04	−.08	.12	−.31	.20	.11	.11	.06	.32	.35					
X_9	.38	.37	.20	.17	.73	−.25	.94	−.22	.22	.04	.09				
X_{10}	−.29	−.30	−.17	.05	.05	.87	−.06	.96	.10	.01	.04	−.19			
X_{11}	−.16	−.11	−.12	.16	.04	.02	.10	.01	.14	.18	−.14	.09	−.04		
X_{12}	.61	.65	.64	−.09	.16	−.17	.16	−.24	−.05	.10	.18	.19	−.24	−.01	
Z	.16	—	.08	.04	.02	−.10	.13	−.09	−.07	−.05	−.17	.10	−.09	.10	.16

[a] U_1 is the target area mean, U_2 is the target area mean of transformed data, covariates are defined in (1.1), Z is the seeding indicator variable.

Table A-II Analysis of variance tables for Models (5)-(9) for the various target areas, transformed data.

Model	Source of Variation	d.f.	Mean Squares Target Areas					F-Ratios Target Areas				
			(i)	(ii)	(iii)	(iv)	(v)	(i)	(ii)	(iii)	(iv)	(v)
(5)	Seeding	1	.57	.74	.00	.24	.46	.05	.21	.00	.02	.06
	Covariates	8	382.54	139.56	278.71	351.30	279.66	30.79	40.07	23.66	31.67	34.38
	Residual	96	12.42	3.48	11.78	11.09	8.13	—	—	—	—	—
(6) Without X_c	Seeding	1	.00	.20	.28	.03	.02	.00	.03	.02	.00	.00
	Covariates	7	374.40	119.69	289.68	349.67	240.59	22.24	18.92	21.09	23.76	17.49
	Residual	97	16.83	6.33	13.74	14.72	13.76	—	—	—	—	—
(7) Without X_{12}	Seeding	1	11.64	3.68	7.89	10.01	4.79	.78	.92	.55	.73	.53
	Covariates	7	398.63	151.58	279.56	362.30	305.04	26.64	38.01	23.73	26.44	33.68
	Residual	97	14.96	3.99	14.39	13.71	9.06	—	—	—	—	—
(8) Without X_c, X_{12}	Seeding	1	44.68	17.04	29.59	39.51	29.52	1.46	1.51	1.23	1.43	1.28
	Covariates	6	200.39	54.18	164.64	188.43	120.14	6.53	4.79	6.85	6.83	5.19
	Residual	98	30.68	11.32	24.04	27.61	23.16	—	—	—	—	—
(9)	Seeding	1	1.56	6.29	.36	.25	11.88	.13	1.82	.03	.02	1.48
	Interactions	7	16.95	3.00	15.00	14.78	8.00	1.41	.87	1.31	1.37	1.00
	Covariates	8	382.54	139.56	278.71	351.30	279.66	31.73	40.35	24.27	32.52	34.89
	Residual	89	12.06	3.46	11.48	10.80	8.02	—	—	—	—	—

Table A-III Regression coefficients for Model (5) for the various target areas, transformed data.

	Covariates	Target Areas				
		(i)	(ii)	(iii)	(iv)	(v)
Constant		.8952	1.6289	.8241	.8758	1.5273
X_c	Control Area	1.6467^b	2.6051^b	1.3451^b	1.5419^b	2.4728^b
X_2	700 mb Spd.	$.0136^b$	$.0105^b$	$.0161^b$	$.0140^b$	$.0103^b$
X_3	700 mb Wind Dir.	$-.0040^b$	$-.0038^b$	$-.0047^b$	$-.0042^b$	$-.0040^a$
X_6	500 mb Temp.	$-.0077$	$.0082$	$-.0143$	$-.0084$	$.0029$
X_7	Stab. Class	$-.1327^a$	$-.1249$	$-.1593^a$	$-.1383^a$	$-.1258$
X_8	Showalter Index	$-.0326^a$	$-.0402^b$	$-.0336^a$	$-.0319^a$	$-.0370^b$
X_{11}	Instab. Transpt.	$-.0001$	$.0000$	$-.0002$	$-.0001$	$.0000$
X_{12}	Duration	$.0042^b$	$.0037^b$	$.0051^b$	$.0043^b$	$.0034^b$
Z	Seeding	$.0164$	$.0373$	$.0008$	$.0111$	$.0200$

[a]Significant at level 0.05.
[b]Significant at level 0.01.

Table A-IV Analysis of variance tables for the various target areas, storms as blocks, transformed data.

Source of Variation	d.f.	Mean Squares Target Areas					F-Ratios Target Areas				
		(i)	(ii)	(iii)	(iv)	(v)	(i)	(ii)	(iii)	(iv)	(v)
Seeding	1	23.97	12.57	12.57	19.80	21.15	0.82	1.14	0.57	0.75	0.95
Storms	37	61.08	18.86	50.66	56.68	40.75	2.08	1.70	2.30	2.16	1.83
Residual	67	29.39	11.07	21.99	26.25	22.24	—	—	—	—	—
R^2		—	—	—	—	—	0.54	0.49	0.56	0.55	0.51

BIBLIOGRAPHY

Bradley, R. A., Srivastava, S. S., and Lanzdorf, A. (1977a). Data summarization in a weather modification experiment: I. A response surface approach. Tech. Report M417, Department of Statistics, Florida State Univ., Tallahassee, FL.

Bradley, R. A., Srivastava, S. S., and Lanzdorf, A. (1977b). Summarization of precipitation data for a weather modification experiment. *Fifth Conf. Prob. and Statist., Preprint Vol.*, 201-205, Amer. Meteor. Soc., Boston, MA.

Bradley, R. A., Srivastava, S. S., and Lanzdorf, A. (1978). An examination of the effects of cloud seeding in Phase I of the Santa Barbara Convective Band Seeding Test Program. Report M467, Department of Statistics, Florida State Univ., Tallahassee, FL.

Brown, K. J. and Elliott, R. D. (1972). Mesoscale changes in the atmosphere due to convective band seeding. *Third Conf. Weather Modif., Preprint Vol.*, 313-320, Amer. Meteor. Soc., Boston, MA.

Brown, K. J., Elliott, R. D., Thompson, J. R., St. Amand, P., and Elliott, S. D. Jr. (1974). The seeding of convective bands. *Fourth Conf. Weather Modif., Preprint Vol.*, 7-12, Amer. Meteor. Soc., Boston, MA.

Brown, K. J., Thompson, J. R., and Elliott, R. D. (1975). Large scale effects of cloud seeding, final report, 1970-74 seasons. Tech. Report 75-2, Aerometric Research Inc., Goleta, CA.

Elliott, R. D. and Brown, K. J. (1971). The Santa Barbara II project — downwind effects. *Int. Conf. on Weather Mod., Preprint Vol.*, 179-184, Canberra, Sept. 6-11, 1971.

Elliott, R. D., St. Amand, P., and Thompson, J. R. (1971). Santa Barbara pyrotechnic cloud seeding test results 1967-1970. *J. Applied Meteorology 10*, 785-795.

Elliott, R. D. and Thompson, J. R. (1968a). Santa Barbara pyrotechnic seeding device program, 1967-68 winter season, final report. Tech. Publ. 4645, Naval Weapons Center, China Lake, CA.

Elliott, R. D. and Thompson, J. R. (1968b). Persistence of nuclei in the Santa Barbara test area, 1967-68 winter season, final report. Tech. Publ. 4646, Naval Weapons Center, China Lake, CA.

Elliott, R. D. and Thompson, J. R. (1969). Santa Barbara pyrotechnic seeding device program, 1967-68 and 1968-69 seasons. Tech. Publ. 4816, Naval Weapons Center, China Lake, CA.

Elliott, R. D. and Thompson, J. R. (1972). Santa Barbara convective seeding test program, 1970-71 season and 1967-71 summary. Tech. Publ. 5308, Naval Weapons Center, China Lake, CA.

Gleeson, T. A. (1977). Data summarization in a weather modification experiment: II. Concomitant variables. Tech. Report M419, Department of Statistics, Florida State Univ., Tallahassee, FL.

Scott, E. (1978). Data summarization in a weather modificaton experiment: III. A multivariate analysis. Tech. Report M442, Department of Statistics, Florida State Univ., Tallahassee, FL.

Statistical Task Force to the Weather Modification Advisory Board (1978). The Management of Weather Resources, Volume II: The Role of Statistics, Report to the Secretary of Commerce, Washington, U.S. Government Printing Office.

Thompson, J. R., Brown, K. J., and Elliott, R. D. (1975). Santa Barbara convective band seeding test program, final report. Tech. Publ. 5804, Naval Weapons Center, China Lake, CA.

Weather Modification Advisory Board (1978). The Management of Weather Resources, Volume I: Proposals for a National Policy and Programs, Report to the Secretary of Commerce, Washington, U.S. Government Printing Office.

Comparing the Testing of Hypotheses Based on Lognormal and Gamma Distributions

M. Hanson and L. Barker

Department of Statistics
The Florida State University
Tallahassee, Florida

Abstract

Tests based on lognormal and gamma distribution for the ratio of the mean of seeded and unseeded rainfall are applied to the data from the Phase II Santa Barbara weather modification experiment. It is shown that these different distributional assumptions lead to very different conclusions, although these two distributions are often considered as alternative contenders in describing rainfall distribution.

1 INTRODUCTION

There appears from time to time discussion of the relative merits of various probability distributions, particularly the gamma distribution and the lognormal distribution, as representatives of rainfall distribution. The fact is clear that there is no compelling experimental evidence in favor of any particular well known distribution. Indeed the extreme variability of rainfall problems both in space and time, make it appear rather unlikely that a general description of rainfall can be embraced by any particular simple family of distributions. This is further complicated in weather modification experiments by the fact, which is sometimes overlooked by hypothesis tests, that the relevant rainfall distribution is highly conditional upon certain factors predetermined, sometimes not very explicitly, by the experimenters, such as there being a certain probability of a certain amount of rainfall without seeding, or that certain meteorological parameters, whose relationship to rainfall may not be very specific, attain certain values before it is decided that a suitable experimental situation exists.

Moreover, in comparing probability distributions as potential models for use in hypothesis testing one should not ignore the purpose for which the experiment is being conducted. In cloud seeding for example one is usually interested in obtaining larger rainfall. Since most of the overall rainfall usually is the result of relatively few storms, it is the tail of the distribution, and not the entire distribution or even perhaps a large part of it that is relevant. So in such a situation, fitting distributions to data should be replaced by fitting tails of distributions to data. Then of course the question arises of how much of the tails is relevant.

These and other questions make it undoubtedly desirable to use nonparametric tests of hypothesis. But nevertheless tests of hypothesis based on distributional assumptions are sometimes used in weather modification experiments; and one very unfortunate theoretical problem is that in the comparison of the relative efficiencies of nonparametric tests it is a deficiency of existing statistical theory that in general it is necessary to make distributional assumptions. It was in connection with the latter problem that we decided to investigate the following tests of hypothesis based upon lognormal and gamma distributions using data from the Phase I Santa Barbara weather modification experiment.

2 TESTS OF HYPOTHESIS

We use the methods discussed by Crow (1977) for determining confidence intervals for the ratio of the means of seeded and unseeded rainfall assuming lognormal and gamma distributions.

Let $X_1, X_2, \ldots, X_m$ and $Y_1, Y_2, \ldots, Y_n$ be independent random samples of two-parameter lognormal variates X and Y; so the logarithms of X and Y have the distribution $N(\mu, \sigma^2)$ and $N(\mu', \sigma^2)$ respectively, assuming equal variances. Let

$$U_i = \ln X_i, \quad V_j = \ln Y_j$$

$$\bar{U} = \frac{1}{m} \sum_{i=1}^{m} U_i, \quad \bar{V} = \frac{1}{n} \sum_{j=1}^{n} V_j$$

$$S_u^2 = \frac{1}{m-1} \sum_{i=1}^{m} (U_i - \bar{U})^2, \quad S_v^2 = \frac{1}{n-1} \sum_{j=1}^{n} (V_i - \bar{V})^2$$

$$S^2 = \frac{1}{m+n-2} [(m-1) S_u^2 + (n-1) S_v^2]$$

Then confidence intervals for the ratio of means $\rho = EY/EX$ are obtained by transforming the confidence intervals for $\mu' - \mu$ using the usual t distribution.

If X and Y are gamma distributed with scale parameters θ and θ' and slope parameters γ and γ', respectively, the probability density function of X is

$$g(x; \theta, \gamma) = \frac{1}{\theta \Gamma(\gamma)} \left(\frac{x}{\theta}\right)^{\gamma-1} e^{-\frac{x}{\theta}}, \quad x > 0$$

and of Y is similarly $g(y; \theta', \gamma')$.

Then a $100(1 - 2\alpha)\%$ central confidence interval for ρ is given by

$$\frac{\bar{Y}}{\bar{X}} \frac{1}{F_{2n\gamma', 2m\gamma, \alpha}} < \rho < \frac{\bar{Y}}{\bar{X}} F_{2m\gamma, 2n\gamma', \alpha}$$

3 RESULT

The tests were applied to the rainfall data at each of 178 raingauge locations. These of course do not represent independent experiments. At the 5% one-sided level of significance it was found that under the lognormal assumption the number of raingauge locations at which seeding significantly increased rainfall was 32, that is, 18%, and under the gamma assumption the number was 28, that is, 16%. However the significant locations were not identical in the two cases. Only 12, that is, 7% of the raingauge locations were significant in common. So although the two distributions are often regarded as contenders in describing the same thing, rainfall distribution, they can lead to very different conclusions in hypothesis testing in a single weather modification experiment. In such a context we feel that it is pointless to debate the merits of either distribution in connection with modification.

Rather than adhering to a 5% significance level we have machine-plotted in Figures 1 (see p. 58) and 2 (see p. 59) contours of equal P values on the geographic region around Santa Barbara for the two distributions, which show more dramatically the differing conclusions resulting from the different assumptions.

BIBLIOGRAPHY

Crow, E. L. (1977). Minimum variance unbiased estimators of the ratio of means of two lognormal variates and of two gamma variates. *Commun. Statist.-Theor. Meth.*, *A6(10)*, 967-975.

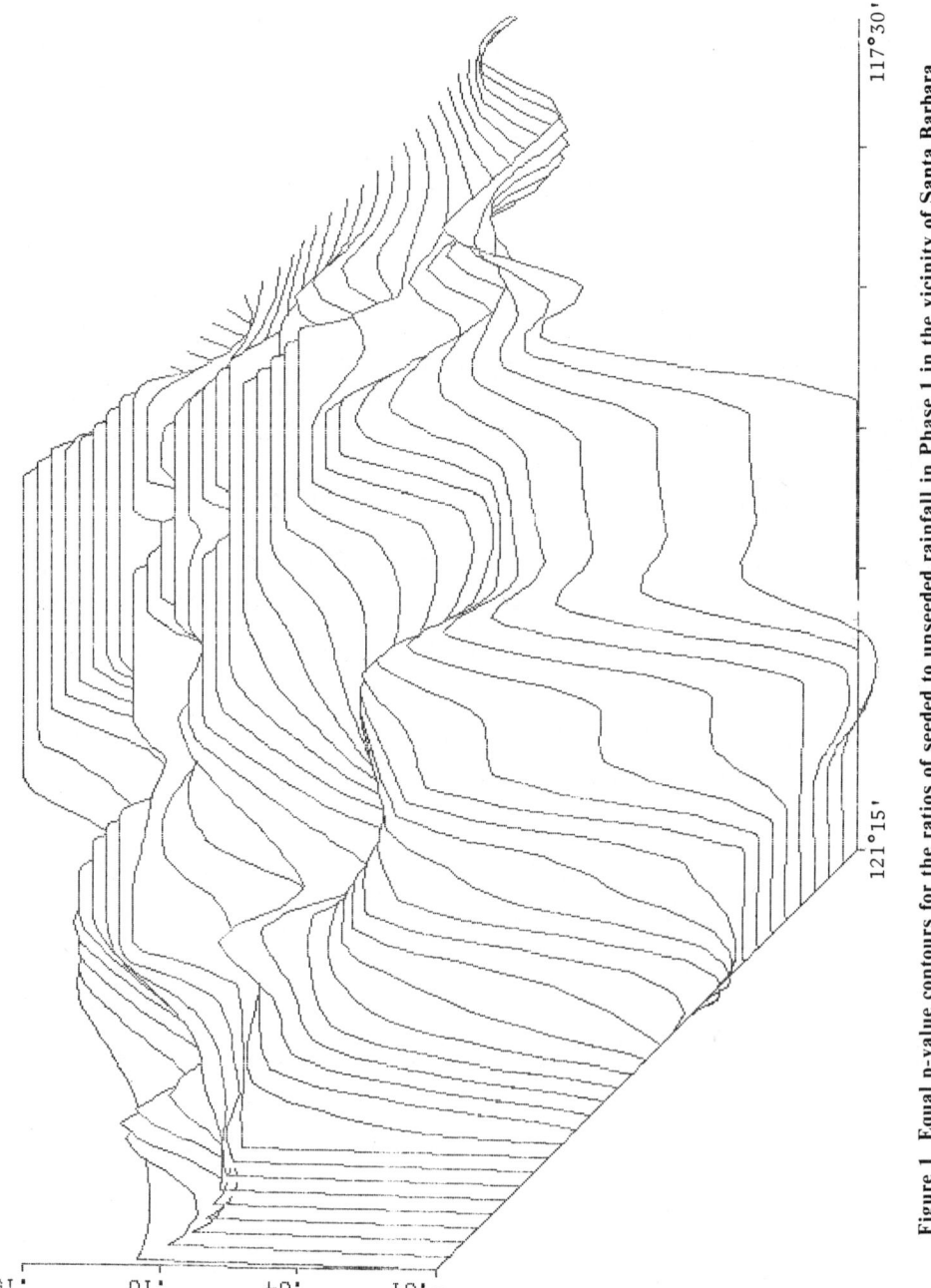

Figure 1 Equal p-value contours for the ratios of seeded to unseeded rainfall in Phase 1 in the vicinity of Santa Barbara (p-valued truncated at .15) assuming a gamma distribution.

Lognormal and Gamma Testing

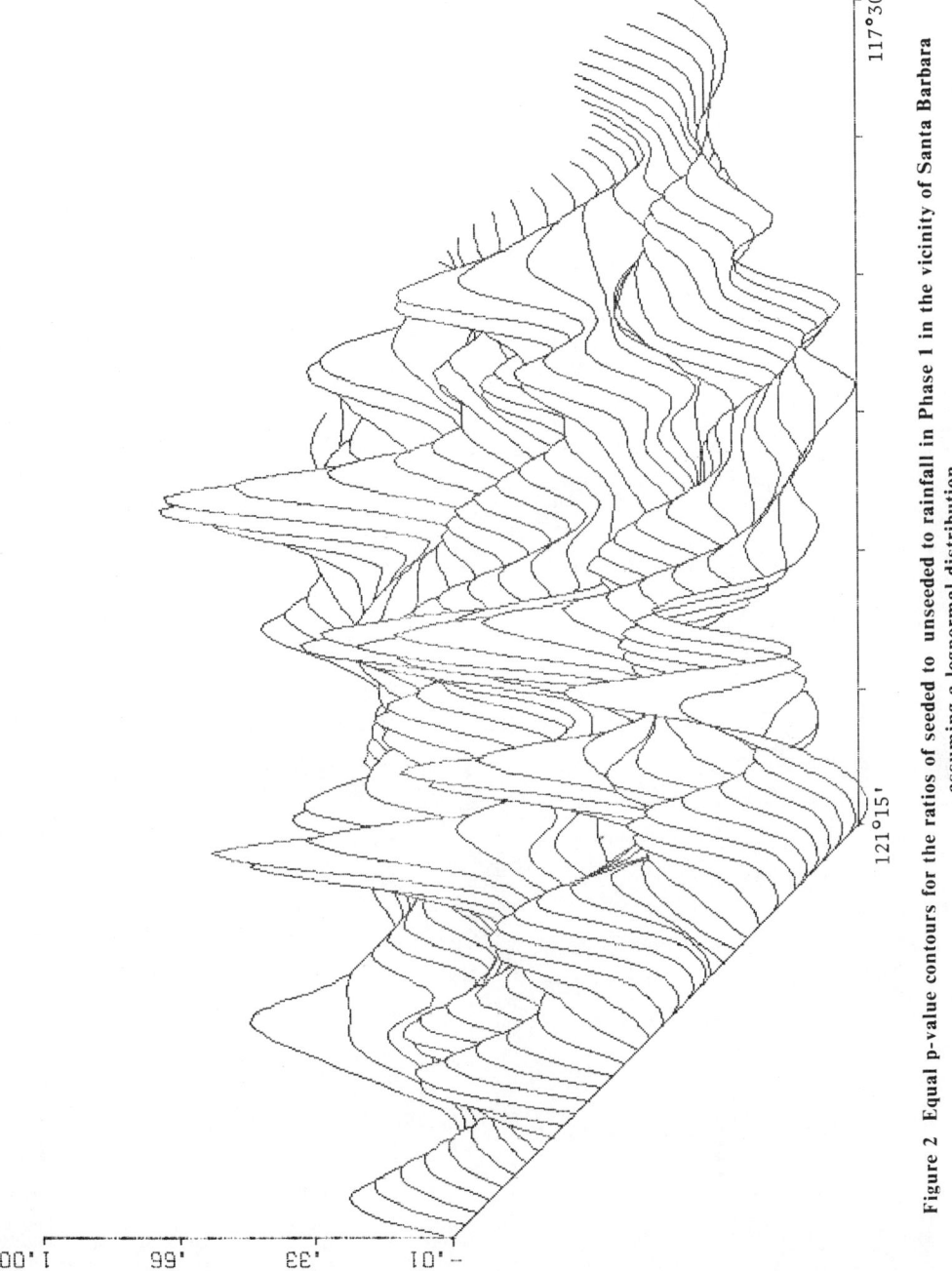

Figure 2 Equal p-value contours for the ratios of seeded to unseeded to rainfall in Phase 1 in the vicinity of Santa Barbara assuming a lognormal distribution.

A Multivariate Methodology for the Analysis of Weather Modification Experiments

Elton Scott

Department of Statistics
The Florida State University
Tallahassee, Florida

Abstract

This paper develops applications of multivariate statistical models, particularly principal component analysis, to the analysis of data from weather modification experiments. The efficacy of these multivariate applications is examined by applying the proposed models to data from Phase I (1967-71) of the Santa Barbara Convective Band Seeding Program conducted for the Navy by North American Weather Consultants. Multivariate summary measures of precipitation are developed and multivariate methods are given to analyze the effects of cloud-seeding on precipitation. Results from these models, based on the above-mentioned data set, are reported along with conclusions and suggestions for further work.

1 INTRODUCTION

Weather modification experiments generally produce multivariate data. Both the precipitation measurements and concomitant variables for a given experimental unit are usually represented as a vector of measurements, and as such, multivariate methods are appropriate for the analysis of the data. Logically, the data analysis can proceed in two steps. First, the precipitation measurements must be represented by an appropriate summary measure and, second, these summary measures for the experimental units can be used to examine the effects of seeding.

Below, in Sections (2) and (3), models for these two stages are reviewed and, at each stage, the models are applied to the Phase I data from the Santa Barbara Convective Seeding Program (SBA-I data hereafter) (Thompson, Brown and Elliott, 1975). The final section of the paper gives conclusions on the analyses, limitations of this application, and suggestions for further work.

Elton Scott is permanently with the Finance Department, School of Business at The University of Florida, Tallahassee.

2 SUMMARY MEASURES OF PRECIPITATION

With most weather modification experiments, a large network of raingages is set up over designated areas to collect precipitation measurements. Often measurements from each experimental unit are simply averaged and the simple average is taken to represent the volume of precipitation on the area for that experimental unit. Inasmuch as the raingages are irregularly spaced, the measurements include some noise, and the measurements tend to be positively (but imperfectly) correlated, the adequacy of such averages to represent the full data set can be questioned.

This section provides a brief intuitive description of a multivariate statistical model for dealing with correlated data sets, Principal Components Analysis (PCA hereafter). A method that uses PCA to produce multivariate summary measures of precipitation data from the experimental units is presented and other uses of PCA are considered. This section closes with an evaluation of results obtained when these methods were applied to the SBA-I data. The measures produced at the first stage can be used in the second stage to examine the effects of seeding and the second stage is considered in Section (3).

2.1 The Concept of PCA

The data array is arranged so that the vectors of precipitation measurements for the experimental units (convective bands) are regarded as the observation vectors and the measurements at raingage stations as (possibly) correlated variables. Let $y_{i\alpha}$ be the measurement at Station i in the αth experimental unit. We use

$$z_{i\alpha} = (y_{i\alpha} - \bar{y}_i)/s_i \quad i = 1, 2, \ldots, p; \alpha = 1, 2, \ldots, N,$$

to calculate the $p \times p$ correlation matrix, $\mathbf{R} = \left[\sum_{\alpha=1}^{N} z_{i\alpha} z_{j\alpha}/N \right]$, where $\bar{y}_i$ is the mean over the N observations at Station i, and s_i is the corresponding estimated standard deviation calculated from the N values (observed at Station i for the N convective bands).

The principal components to be determined depend on the characteristic roots and vectors of $\mathbf{R}$. Let the kth characteristic root and corresponding characteristic vector be given by λ_k and $\mathbf{a}_k$, where elements of $\mathbf{a}_k$ are a_{ki}, $i = 1, 2, \ldots, p$, $k = 1, 2, \ldots, p$. The kth principal component for the αth experimental unit is:

$$c_{k\alpha} = \sum_{i=1}^{p} a_{ki} z_{i\alpha}.$$

Of course the $\mathbf{a}_k$ and λ_k must be estimated and these estimates (designated as $\hat{\mathbf{a}}_k$ and $\hat{\lambda}_k$ below) are used to estimate the principal component values for each observation vector.

The estimated product-moment correlation between the ith variable and the kth component is given by $\hat{a}_{ki} \sqrt{\hat{\lambda}_k}$ (Morrison, 1976, p. 271). Thus, coefficients estimated from standardized data (as above) indicate the extent and sign of the association between variables (stations) and principal components.

The principal components are ordered by the proportions of total variance represented by them. If the characteristic roots are extracted from a correlation matrix, as in the present application, the sum of the characteristic roots will be p and the proportion of the total variance attributable to the kth component will be λ_k/p.

Since PCA is applied so that the dimensionality of response vectors can be reduced, one would like the first few principal components to account for most of the variance of the data. Thus, if J of the components "adequately" represent the data, the p-dimensional variate vector could be summarized by the corresponding J-dimensional principal component vector.

All variables usually have high pairwise correlations with the first principal component and further, the second and subsequent components may exhibit strong associations with subsets of variables. If the first few principal components account for most of the variance, and if each principal component is highly correlated with one or more representative stations, future data collection might be simplified. Data from these representative stations should essentially contain the information of the principal components with which they are associated, and, since the components represent the essential information from the full network of stations, most of the information from the full network should be reflected in the information from the representative subset. If these assumptions hold, the efficiency of future data collection could be improved since costs should be reduced by improvement in the reliability of data collection at the subset of stations and collection of data from the representative subset of stations only, rather than from the full network.

2.2 Application of PCA to Weather Modification Data

The application of PCA to weather modification experiments would proceed with initial estimation of principal components from available data. Given that the estimates indicate that the first few principal components do account for most of the variance in the full data set, a subset of key stations could be selected to reduce the costs of future data collection in weather modification experiments. If the data are pooled in this manner, we must assume that the principal axes of data are not affected by seeding. That is, the approach assumes that the principal components are the same for both seeded and unseeded experimental units.

PCA can also be applied to evaluate weather modification experiments. The estimates of principal component values for experimental units are vector summary response measures. These vectors can be compared to other summary measures, such as the mean precipitation, to determine the effectiveness of the PCA approach to summarization of the data. If this approach is useful, the vectors for the seeded data could be compared to the vectors from the unseeded data to determine the effects of seeding. This analysis is pursued in Section (3) of this paper. The patterns of the station-component correlations may also prove useful in evaluation of the nature of the precipitation response over an area.

In most applications of PCA, the simple mean of the variables is associated with the first principal component. If the band means and the first principal component scores are highly correlated, then the second and subsequent components must represent essential features of the variates that are not reflected in their mean. For precipitation data, the second and subsequent components could be associated with subarea concentrations of precipitation.

2.3 Evaluation of PCA Results

An obvious question on the above procedure is "How well do the first few components represent the data from the full network of stations?" This question can be answered through use of regression analysis to "predict" summary response measures of interest with the principal component values as the independent variables. Thus, if the mean is the summary measure of interest, it should be reasonably represented as a linear function of the first few principal components.

An alternative evaluation of the usefulness of the PCA approach depends on the ease of the interpretation of the components. The station-component correlations may be interpretable through examination of a graphical presentation of the station-component correlations. Interpretation procedures often are tenuous however, because the principal components sometimes do not present interpretable patterns for the variables.

If the PCA results compare favorably to other summary measures, the usefulness of a key subset of stations in representing the summary measure can also be evaluated from the regression of the summary measure on the key station measurements.

Thus, the steps suggested in application of PCA to precipitation data are:

1. Estimate principal components over the full network of stations.
2. Examine the efficacy of the principal components as a summary response measure.
3. Interpret graphical presentations of the station-component correlations as appropriate.
4. Select a key subset of stations that are highly correlated with the first few principal components.
5. Test the effectiveness of these key stations in representation of the information in the summary measure of interest.

Final steps would then use the above results to test for the effects of seeding. These testing steps are detailed in Section (3) of this paper. The remaining subsections of this section report the results of the above steps for data from Phase I of the Santa Barbara experiments (Thompson, Brown, and Elliott, 1975).

2.4 The Santa Barbara Data

The Santa Barbara experiments and data are briefly reviewed by Bradley, Srivastava, and Lanzdorf (1977) and detailed in Thompson, Brown, and Elliott (1975). The frequency of missing data at many of the stations and the nature of PCA required an adjustment in the set of stations used in this analysis. The basic areas of interest are the "Control" and "Target" areas (see Bradley, Srivastava, and Lanzdorf, 1977), but the number of stations in each area was reduced considerably because of missing data.

The control and target areas were analyzed separately to be consistent with the response surface approach of Bradley, Srivastava, and Lanzdorf (1977). The resulting principal component values could be used as covariates or summary measures in subsequent analyses.

As mentioned above, many stations had large numbers of missing measurements (a complete set of measurements for a station would include precipitation measurements for all 107 Phase I bands). Stations with large numbers of missing values present the possibility of a sample covariance matrix that is not positive definite. When PCA was applied to all stations, irrespective of missing data, sizeable negative characteristic roots occurred, especially for the target area. To avoid this problem, all stations with more than 22 missing band values were excluded from the analysis (this cutoff point is arbitrary, but it did improve the behavior of the latent roots). This deletion procedure yielded nineteen control stations (fifteen were deleted) and seventy-two target area stations (thirty-eight were deleted). A complete list of included stations is given by Scott (1978).

The elimination of stations could have systematically excluded some geographic subareas in either target or control areas. To test for this possibility, multivariate analysis of variance (MANOVA) was applied to the vector of location coordinates (latitude, longitude, and altitude). MANOVA is the multivariate analog of univariate ANOVA (Morrison, 1976). The "treatment" groups were the included and excluded sets of stations, and the null hypothesis was that there were no differences in the mean location vectors for the two groups.

The null hypothesis could not be rejected for the control-area groups but for the target-area groups, the null was rejected ($\alpha = .05$). This implies that there are significant differences in the coordinates of the excluded and included stations. This could produce systematic variations for results on the included subset relative to results for the full network of stations.

The excluded stations for the target area were primarily in the northeast quadrant of the target area, so they are remote relative to the seeding site. Because these stations are generally remote, the systematic exclusion of these stations should not limit the extent to which results below can be generalized. In any event, the advantages of including stations with large amounts of missing data were not considered to be important enough to overcome the problems produced by including these stations.

2.5 Principal Component Estimates

As specified above, the data were standardized and correlation matrices were estimated for the stations over the 107 bands. The pairwise simple correlations estimates on **R** were based on all data available in common for each pair of stations. Characteristic roots and associated characteristic vectors were obtained separately for the 72-square correlation matrix for the target area and for the 19-square correlation matrix for the control area and corresponding values for the principal components were obtained for each convective band. Table 1 gives the characteristic roots, proportions of variance, and cumulative proportions of variance that are accounted for by the first few principal components of both target and control areas. These results suggest that the first few principal components do indeed account for much of the variance. Thus, a few dimensions of the principal component vector accounts for most of the variance in the precipitation measurement vectors.

Table 1 Eigenvalues and percent variance associated with primary principal components.

Principal Component Number	Eigenvalue	% of Var.	Cum. % Var.
Target Area (72 Stations):			
1	51.33	71.3	71.3
2	4.80	6.7	78.0
3	4.23	5.9	83.8
4	2.07	2.9	86.7
5	1.39	1.9	88.7
Control Area (19 Stations):			
1	14.45	76.1	76.1
2	1.27	6.7	82.8
3	.89	4.7	87.4
4	.59	3.1	90.5
5	.37	1.9	92.4

2.6 Response Information from Principal Components

The information content of the first few principal components of the data is analyzed below. Separate band precipitation means (and band precipitation standard deviations) were estimated from the stations included in the PCA data over target and control areas. These values were used as dependent variables and were regressed on the first few principal component values for the target and control areas (The first six principal components for the target area and the first five principal components for the control area). Generally means and standard deviations of precipitation measurements for experimental units are positively (but not perfectly) correlated. It is possible though, for cloud seeding to affect the scale parameter of the distribution without affecting the location parameter, so standard deviations were also included as dependent variables in similar regressions. The linear regression results are reported in Table 2.

Table 2 Regression results predicting means and standard deviations from principal component vectors.

Dependent Variable	Adj. R^2	F (df_1, df_2)	MSE	Const.	PCS1	PCS2	PCS3	PCS4	PCS5	PCS6
Target Area:										
Mean	.998	7,987 (6, 100)	.000	.251	.283	.002	.004	−.000	−.003	−.005
Std. Dev.	.888	141.1 (6, 100)	.002	.167	.145	.021	.015	.002	−.020	−.004
Control Area:										
Mean	.989	1,900 (5, 101)	.000	.242	.197	.006	−.002	.007	.006	—
Std. Dev.	.689	47.5 (5, 101)	.002	.117	.058	.025	−.006	−.010	.012	—

The regression results confirm that the first few components yield near-perfect predictions of the means. Although it is not reported in Table 2, the first principal component alone accounted for most of the variance in band means. The R^2 for simple regressions of the means on the first principal component were .997 for the target area and .985 for the control area. Thus the first principal component generally reflects the summary information in the band means and the second and other principal components reflect further summary response information.

Values of R^2 for the standard deviations are lower but the association with the first few principal components is still quite strong. Here, the first principal component is again strongly associated with the standard deviations but other components are also important.

The statistics given in Table 2 demonstrate the efficacy of the principal components as summary response measures for precipitation. Response-surface integrated volumes were not directly considered because these values are highly correlated with the simple band means (Bradley, Srivastava, and Lanzdorf, 1977).

2.7 Interpreting of the Principal Components

As mentioned above, principal components are not always interpretable as recognizable constructs. In the present case however, the results do yield readily interpretable results. The station-component correlation estimates ($r_{ki} = \hat{\alpha}_{ki}\sqrt{\hat{\lambda}_k}$) for the first three components of both target and control areas are given in Figures 1 through 6. The left side of the horizontal bars represent station locations in degrees and hundredths and the vertical spikes represent the

r_{ki} or the simple correlation between the kth component and the ith station (the scale is 2 inches equals a correlation of 1.0). Correlations between station precipitations and the first component (Figures 1 and 4) are large for all stations in both areas. For the second component (Figures 2 and 5), the correlations are associated with the location of the stations relative to the coast. The signs are opposite for the inland stations relative to the coastal stations. Given the topography of the Santa Barbara area (i.e., mountain ridges along the coast), this component is interpreted as an orographic component of precipitation. This pattern holds for both target and control areas, as does the third component pattern (Figures 3 and 6). The magnitudes and signs of the correlations for the third component vary directly with the east-west location of the stations. Again, the pattern is clearly present for both target and control areas. The patterns for the other principal components are not clearly interpretable and are not displayed, but these components are relatively unimportant since most of the variance is associated with the first three components.

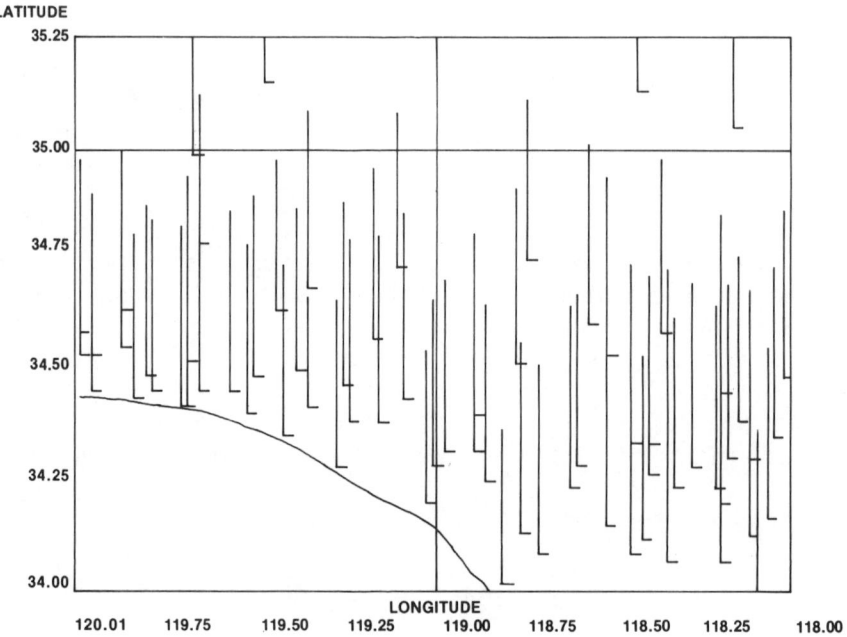

Figure 1 Correlations between precipitation and principal component 1 by stations, target area.

A Multivariate Methodology for Analysis

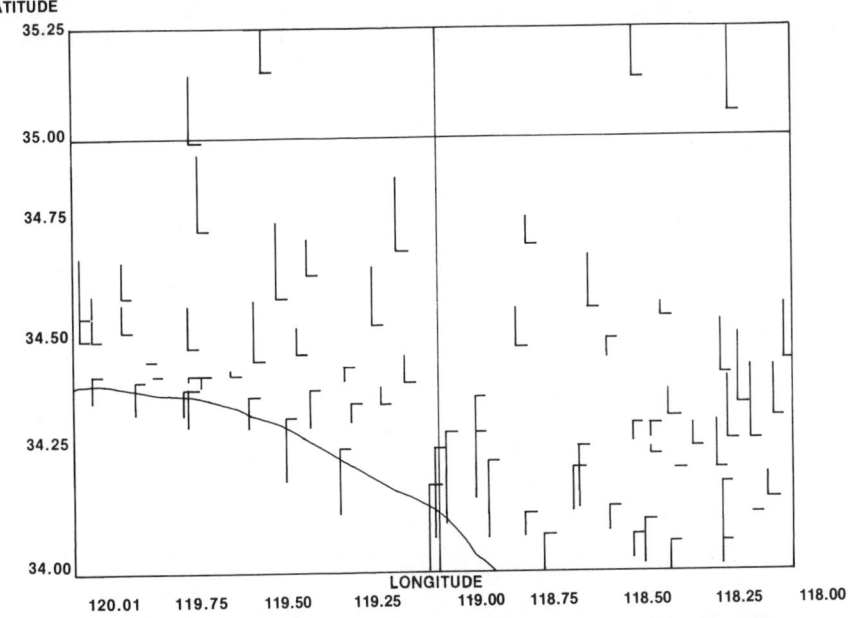

Figure 2 Correlations between precipitation and principal component 2 by stations, target area.

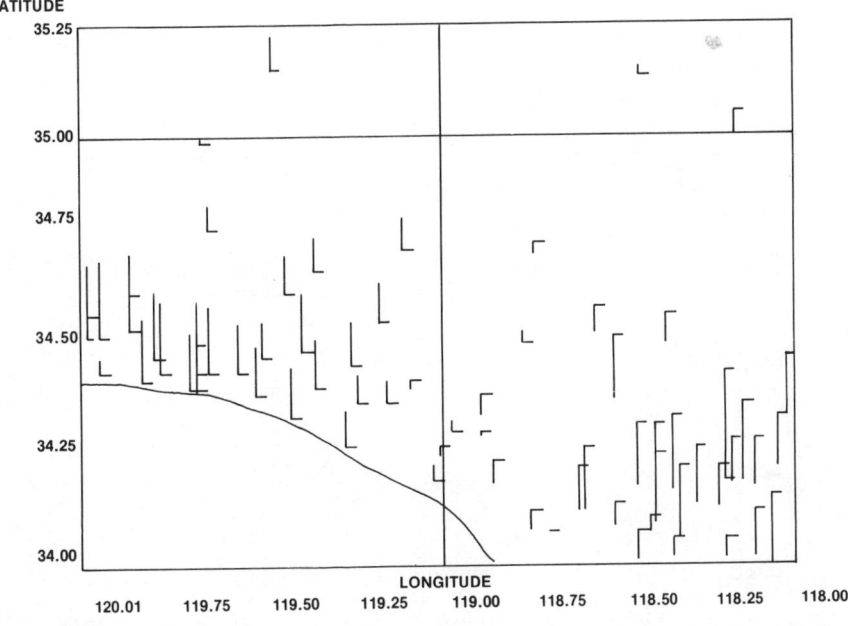

Figure 3 Correlations between precipitation and principal component 3 by stations, target area.

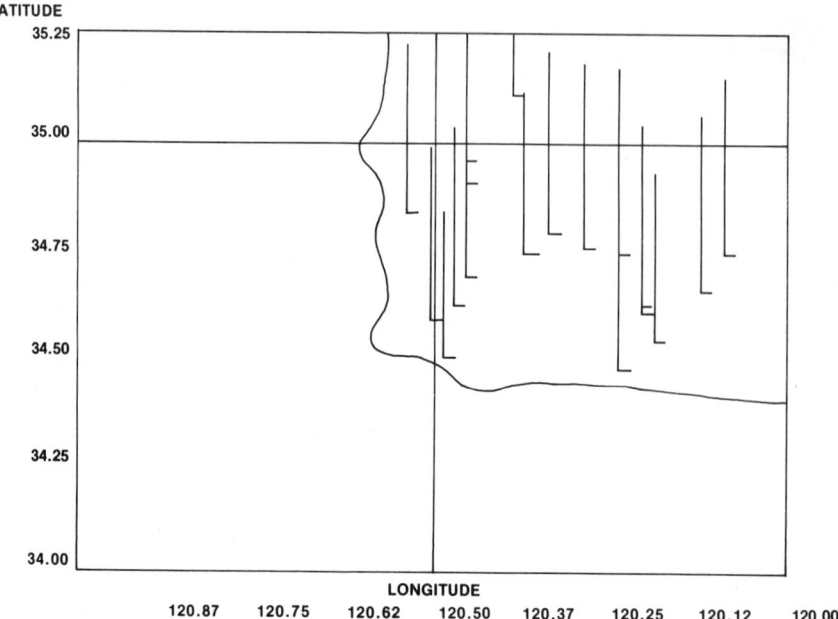

Figure 4 Correlations between precipitation and principal component 1 by stations, control area.

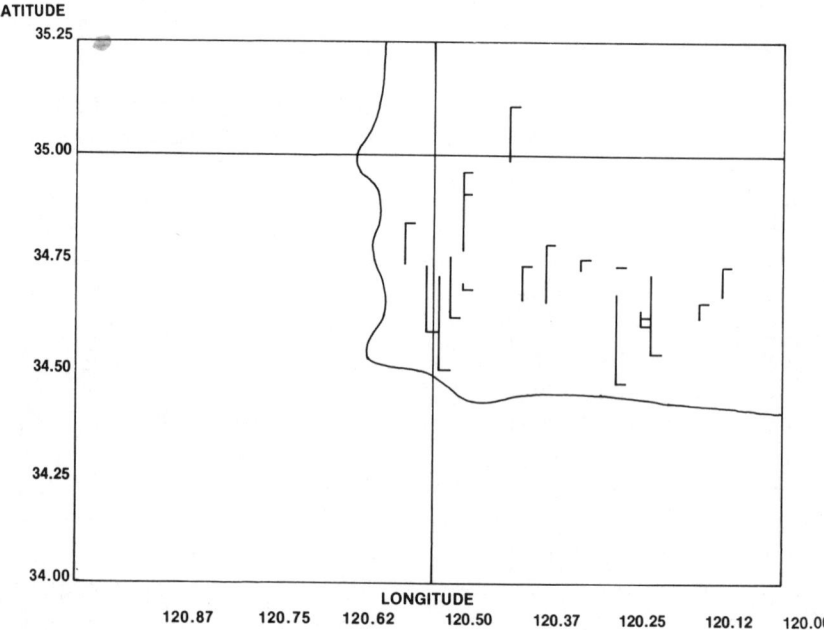

Figure 5 Correlations between precipitation and principal component 2 by stations, control area.

A Multivariate Methodology for Analysis

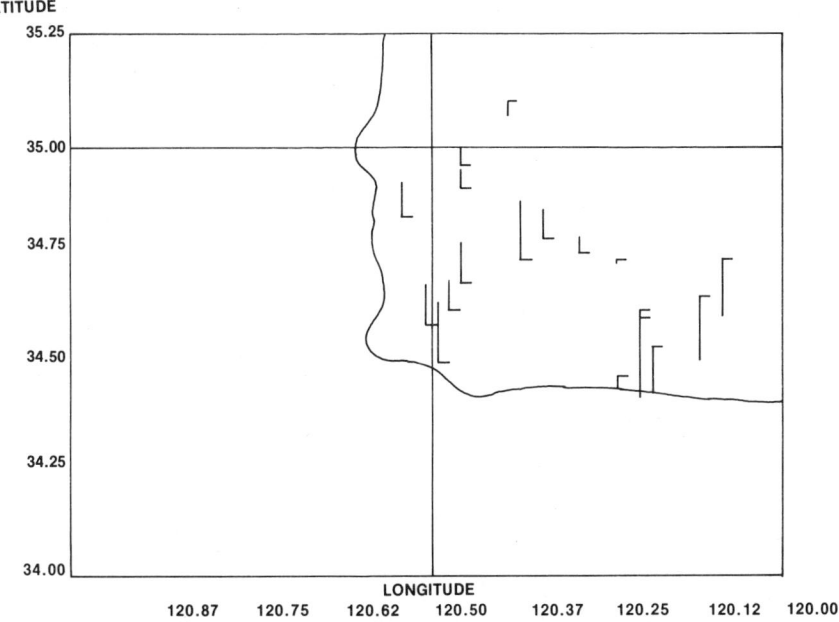

Figure 6 Correlations between precipitation and principal component 3 by stations, control area.

2.8 Selecting Key Stations

The final part of this section reports on the use of PCA to select a small subset of key stations that can be used to reproduce summary response measures.

A subset of stations that were highly correlated with the second through the sixth components were selected (The first component was not considered because it is highly correlated with most stations). For the target area, two stations were selected for each of the five components and for the control area only one station was selected for each component. Two other objectives were considered in addition to maximization of the correlation for the target area stations: (i) Stations with fewer missing values were chosen where several stations had nearly-equal correlations: (ii) For the second station in the target area, the station was also chosen to be geographically separated from the first station. The selected stations are given in Table 3 along with station-component correlations. In both the target and control stations, some stations were highly correlated with two of the components. This resulted in only eight key target area stations and four key control area stations.

The station-component correlations are not the primary concern here. These key station measurement should reproduce the summary response measures if this approach is to be useful. The results for a series of regressions on these key stations are reported in Table 4.

For the target regressions, the values of the simple correlation coefficients are good (but not outstanding) for the means, standard deviations, and the first two principal components. This suggests that, if means are chosen as the sum-

Table 3 Stations selected to represent principal components.

Key St. Number	Name (SBA No.)	Lat.	Long.	Alt. (m)	No. Miss. Bands	Correlations with PC No.:[a]					
						1	2	3	4	5	6
Target:											
KEYTG 1	E3751(29)	34.27	118.40	335	0	.877	.030	−.364	−.030	.057	−.111
KEYTG 2	E7735(40)	34.75	118.73	1377	8	.790	.127	−.064	.459[a]	.037	−.225[a]
KEYTG 3	E8752(47)	35.15	119.47	312	4	.648	.364	.158	.028	.277[a]	.245
KEYTG 4	S211(130)	34.45	119.78	122	3	.858	−.008	.364[a]	.012	.236	−.039
KEYTG 5	S221(134)	34.98	119.67	662	3	.750	.340	.020	.034	.285[a]	.218[a]
KEYTG 6	S238(141)	34.78	119.65	1527	14	.745	.381	.115	.414[a]	−.109	.004
KEYTG 7	V168(154)	34.20	118.20	11	2	.670	−.480[a]	.217	.289	−.028	−.147
KEYTG 8	V190(164)	34.28	119.00	168	4	.841	−.447[a]	.050	.074	−.093	.066
Control:											
KEYCN 1	E4144(30)	35.01	120.38	218	10	.823	−.274	−.068	.065	.302[a]	.340[a]
KEYCN 2	NO1(103)	34.83	120.53	91	12	.848	−.203	.177	.319[a]	−.123	−.094
KEYCN 3	NO4(106)	34.60	120.20	104	4	.891	.034	−.401[a]	.407	.095	−.099
KEYCN 4	S206(127)	34.47	120.23	9	7	.804	.444[a]	−.064	.021	−.112	.038

[a]The highest correlations (the primary basis for selecting these stations) are indicated for the stations.

Table 4 Regressions on Key Stations

Sum. Stat. and Ind. Var.	Mean	Std. Dev.	PCS1[a]	PCS2[a]	PCS3[a]	PCS4[a]
Target:						
Adj. R^2	.902	.735	.908	.710	.574	.403
MSE	.007	.005	.081	.256	.371	.522
F^b	122.3	37.8	130.9	33.5	18.7	9.9
KEYTG 1	.405	.163	1.407	−.107	−3.112	−1.291
KEYTG 2	.002	.046	−.024	.539	−.183	.544
KEYTG 3	−.056	−.044	−.153	.241	−.661	.659
KEYTG 4	.200	.057	.686	−.506	1.549	−1.134
KEYTG 5	.351	.040	1.445	2.252	.606	−2.332
KEYTG 6	.138	.188	.452	2.409	1.145	2.925
KEYTG 7	.104	.041	.416	−1.269	.688	.768
KEYTG 8	.083	.048	.280	−1.052	.229	.551
Constant	.026	.059	−.799	−.090	−.123	−.073
Control:						
Adj. R^2	.646	.359	.674	.214	.174	.126
MSE	.013	.004	.297	.753	.822	.975
F^b	49.4	15.8	55.9	8.2	6.6	4.83
KEYCN 1	.123	.037	.565	−.291	−.044	.472
KEYCN 2	.146	.021	.781	−.347	.924	.649
KEYCN 3	.258	.084	1.443	−.336	−1.173	−.984
KEYCN 4	.129	.059	.576	1.689	−.608	.841
Constant	.127	.076	−.588	−.236	.110	.020

[a]PCSi is the ith Principal Component Score.
[b]All regressions were run on all 107 bands so the F statistic is based on 8 and 98 degrees of freedom for the target regressions and on 4 and 102 degrees of freedom for the control regressions.

mary measure, this approach may be useful in reducing the number of data collection points for large areas. The results on the standard deviation and the second through the sixth component are not adequate for reliable summarization.

It is almost certain that better results could be obtained here by the inclusion of stations that are highly correlated with the first principal component (several station/first-component correlations exceed .95). All stations in the regressions necessarily had lower-than-average correlations with the first principal component because of their higher-than-average correlations with components orthogonal to the first component. Again, this conjecture was not tested here but should be considered in any subsequent work that attempts to use selected stations for the summarization of precipitation. The results on the control area stations are poor enough to indicate that the key station approach is not useful for smaller areas and smaller numbers of stations.

In summary, the key-station approach as used here had limited utility for reducing costs in weather modification experiments. However, for large areas, if the mean is to be used as the summary response measure, the approach could be useful, and probably could be improved on by inclusion of stations with high first-component correlations.

3 MULTIVARIATE ANALYSES OF CLOUD-SEEDING EFFECTS

The models that are covered below do not exhaust the multivariate application in the analysis of data from cloud-seeding experiments. The applications reported here ignore approaches that use concomitant variables because there are indications that the covariates from the SBA-I data were contaminated by the seeding operations (Bradley, Srivastava, and Lanzdorf; 1977). In a forthcoming technical report I have developed alternative methods of analysis (canonical correlation and multivariate analysis of covariance) that incorporate the covariates to reduce error variance in tests of the effects of seeding.

This section considers two methods of analysis that rely on PCA and in each case, the methods are applied to the SBA-I data. This precipitation data obviously violates the assumptions that are necessary for parametric tests of significance, but some of the familiar statistics are produced as descriptive measures.

The multivariate precipitation measurements could respond to seeding in several ways. Cloud-seeding could have no effect, it could affect the means of the vectors of precipitation measurements while leaving the covariances between stations unchanged, it could leave the means unaffected but have an effect on the covariance structure, or both means and covariances could be affected by seeding. If seeding has no effect, then the principal components, and vectors of principal component scores, estimated from all available observations should be essentially the same for the seeded and unseeded experimental units.

If however, seeding affects the means and/or covariances, the analysis should consider summary measures that are estimated separately for the seeded and nonseeded observations. The two subsections below present analyses based on these alternative assumptions.

3.1 Analysis of Principal Components from Pooled-Data Covariance

The discussion and results in Section 2 implicitly assume that the means and the covariance structures are identical for seeded observations and unseeded observations. If this is true, the means of the principal component score vectors should be the same for seeded and unseeded observations. Thus a simple test for the effects of seeding would be to test the hypothesis that the mean vector of principal component scores for seeded observations equals the mean vector of principal component scores for unseeded observations.

It is possible that scores for some components have different means although the vectors of means do not differ significantly. For example, univariate tests of the orographic component scores could indicate differences for seeded and unseeded bands even though a vector test could obscure a significant difference. The subsections below describe methods for the analysis of vectors of scores and for the analysis of univariate scores as well.

A Multivariate Comparison of Principal Component Scores. Hotelling's T^2 statistic is an appropriate test statistic for testing the hypothesis that $\mu_1 = \mu_2$ when the two populations of interest have multivariate normal distributions with a common covariance matrix, Σ of full rank (the location vectors may differ, however). These assumptions do not hold for the principal components scores of precipitation data, especially when the principal components are extracted from the correlation matrix (see Morrison, 1976) for a discussion of the distributional problems here). Nonetheless, if these limitations are recognized as one interprets the results, Hotelling's T^2 statistic can indicate similarities in the vectors of mean principal component scores, since a true null hypothesis would not be rejected although the power to reject a false null hypothesis is limited.

For the principal component results given in Section 2, Hotelling's T^2 was computed to compare the vectors of the means of the first three principal component scores for the Target Area, the computed T^2 value corresponded to an F-ratio of 1.016 with 3 and 103 degrees of freedom (At $\alpha = .05$ the critical value of F is 2.7 and 3 and 100 degrees of freedom). If the distributional assumptions held, this result would indicate negligible effects for seeding.

Univariate Comparisons of Principal Component Scores. As the introduction to this section notes, univariate comparisons could indicate differences for the means of the individual components that get "washed out" in the multivariate statistics. Of course if a series of univariate tests of significance are produced, such multiple-comparison tests suffer from a loss of power. In the present case, the data do not meet the usual assumptions for tests of significance, so the statistics are produced for descriptive purposes only.

The t-statistics (the difference in the estimated mean principal component scores for seeded and nonseeded data, divided by the pooled standard deviations of the principal component scores) for the principal component scores of the target area are given in Table 5. Only the first component means are very different and this indicates that if there are differences in the principal components from the pooled data, the differences seem to be associated with the band means.

A Multivariate Methodology for Analysis

Table 5 Univariate t-statistics for the first three principal component scores from the pooled target area data.

Component	t-Statistics (N = 107 bands)
1	1.67
2	.52
3	.07

This concludes comparisons of the seeded and nonseeded subsets of components based on a pooled-data set. The next subsection presents results obtained when the data was split into seeded and unseeded bands to produce two sets of principal component estimates.

3.2 The Effects of Seeding on the Covariance Structure

Separate PCA estimates (one from the seeded experimental units and one from the unseeded experimental units) can provide insight into the effects, if any, of cloud-seeding on the covariance structure of the precipitation measurements. First, the data from the stations are separated, then standardized in the manner of Section (2.1) to produce separate estimates of correlation matrices for the seeded and unseeded observations. The principal components estimated from these correlation matrices should reflect changes, if any, in the station-to-station covariances that are produced by seeding. Tests for significant differences are again limited since the data does not meet the requisite assumptions for tests of significant differences.

A summary of the results on the estimated eigenvalues for the seeded and unseeded data sets as well as the previously-reported results on the combined data set are given in Table 6. This table shows that the first seeded vs. non-

Table 6 Eigenvalues and percent variance associated with the primary principal components from the seeded-data, unseeded-data and combined data.

Comp. No.	Seeded Data Only (N=56)			Unseeded Data Only (N=51)			Combined Data (N=107)		
	Eigenvalue	% Var.	Cum. % Var.	Eigenvalue	% Var.	Cum. % Var.	Eigenvalue	% Var.	Cum. % Var.
Target Areas (72 Stations)									
1	53.1	73.7	73.7	48.7	67.7	67.7	51.3	71.3	71.3
2	4.9	6.8	80.5	7.4	10.2	77.9	4.8	6.9	78.0
3	4.2	5.8	86.2	4.3	6.0	83.9	4.2	5.9	83.8
4	2.2	3.1	89.3	2.7	3.7	87.6	2.1	2.9	86.7
5	1.3	1.9	91.2	2.0	2.7	90.3	1.4	1.9	88.7
Control Area (19 Stations)									
1	14.7	77.4	77.4	14.4	75.8	75.8	14.5	76.1	76.1
2	1.4	7.5	84.9	1.4	7.4	83.2	1.3	6.7	82.8
3	.8	4.3	89.2	.9	4.9	88.1	.9	4.7	87.4
4	.5	2.8	91.9	.6	3.3	91.4	.6	3.1	90.5
5	.4	1.9	93.9	.3	2.0	93.4	.4	1.9	92.4

seeded eigenvalues differ by 6% of the total variance for the target area versus 1.6% the total variance for the control area. Even when the requisite assumption of multivariate normality holds, the variances of the sampling distributions are so large that the target area difference would be a nonsignificant difference (Press, 1972). The results do indicate that the seeding operations produced precipitation measurements over the target area raingage network that were more correlated with the first component (and thus, the band means) of the precipitation and further, the measurements were less correlated with the second (orographic) component. The same result does not hold for the control area data where the differences were less marked for all components.

The changes in the first two eigenvalues could be accompanied by changes in the patterns of station-component correlations (as in Figures 1-6). Although graphs are not reproduced here because of space considerations, the basic patterns of signs for the first three components were similar to the patterns of signs in Figures 1-6 for both the seeded and nonseeded component estimates. Of course, the relative heights of the spikes (which represent the stations-to-component correlations) changed to reflect increases or decreases in correlations.

In conclusion, the proposed analysis and the results obtained with SBA-I data suggest that analyses of weather modification data should consider the effects of seeding on the covariance structure of the precipitation measurements. The data here represent such gross violations of the requisite assumptions for inference that no conclusions can be reached on the effects of seeding on the covariance structure of precipitation measurements. In general, the nature of the distributions of precipitation measurements precludes tests based on assumptions of normality.

CONCLUSIONS AND IMPLICATIONS

As is the case with many other analyses of weather modification data, the conclusions are limited and the implications for further analysis nearly boundless. The premise that multivariate models can be applied to data from weather modification data is supported by the results here.

The applications of PCA to precipitation measurements produced readily-interpretable principal components. For both control and target areas, the first components were almost perfectly correlated with the simple average of the experimental units, the second component was readily interpreted as an orographic component, while the third component was identified as an east-west component. The use of PCA to select a subset of key stations has utility for large areas but the method produced inadequate summary measures for small areas with few stations.

The use of multivariate methods to analyze the effects of cloud-seeding is also limited. The data, particularly the standardized covariance matrix (i.e., the correlation matrix) fails to meet the multivariate-normality assumption so that the results do not allow the usual inferences as to the effects of cloud-seeding. These problems could probably be overcome with nonparametric methods. The estimates obtained from the SBA-I data indicate that the seeded bands produced precipitation measurements that are uniformly more highly correlated with the band means than the measurements for the unseeded bands.

Concomitant variables were not included in the analysis reported here but work now underway includes these measures in some alternative methods for the analysis of weather modification data. I will report on these results, and on some nonparametric tests of the results reported in this paper, in a forthcoming technical report.

Acknowledgment

This research was supported by the Office of Naval Research under Contract No. N00014-76-C-0394. Reproduction in whole or in part is permitted for any purpose of the United States Government.

BIBLIOGRAPHY

Bradley, R. A., Srivastava, S. S. and Lanzdorf, A. (1977). Data Summarization in a Weather Modification Experiment: I. A Response Surface Approach. ONR Technical Report No. 117.

Morrison, D. F. (1976). *Multivariate Statistical Methods.* Second Edition, New York: McGraw-Hill.

Press, S. J. (1972). *Applied Multivariate Analysis,* New York: Holt, Rinehart and Winston, Inc.

Scott, E. (1978). Data Summarization in a Weather Modification Experiment III: A Multivariate Analysis. ONR Technical Report No. 127.

Thompson, J. R., Brown, K. J., and Elliott, R. D. (1975). *Santa Barbara Convective Band Seeding Test Program, Final Report,* Naval Weapons Center, China Lake, California.

Physically Meaningful Covariates

R.D. Elliott*

North American Weather Consultants
Goleta, California

Abstract

The general problem of evaluating the effects of seeding on the waterbalance over a large area such as a mountain barrier is discussed. The role of models, observational parameters, and covariates is outlined. The stable orographic cloud is chosen for illustrating how systematic measurements of air-mass properties can be used not only as covariates to increase the sensitivity of the statistical comparisons of seeded and not seeded samples, but also as a means for calibrating an orographic conceptual model, thus increasing our understanding of the basic physical mechanisms.

1 INTRODUCTION

Penetrating cloud physics aircraft and weather radar provide a means for observing a "seeding signature" within a seeded cloud, whether it be a microphysical response such as the transformation of liquid droplets to ice particles and thence to precipitation, or a dynamic response such as growth in the depth and vigor of the cloud. The entire treatment/observation cycle can be replicated frequently enough so that careful statistical design and analysis tend to be superfluous. Given an adequate time slot for observation, even the "blind" observer soon learns how to distinguish between treated and untreated clouds.

When a "large area" seeding program is conducted over an entire orographic barrier or over a several hundred square kilometer field of cumulus clouds, with treatment by multiple fixed and/or fast maneuvering sources of ice nuclei, the detection of seeding effects is more complex. There are interactions between various parts of the cloud system that are difficult to observe directly, and in detail, but which are important in the seeding-produced alterations in the area's water balance. In a large area research project it is important to make observations and treatments under a statistical design, using observations of a type that would ensure detection of any seeding-produced changes in the area's water balance, even though some of the more detailed links in the chain of events, the seeding signatures, are not adequately covered.

*R.D. Elliott is currently with NAWC in Salt Lake City, Utah

Natural precipitation and seeding-produced changes therein can be represented by sets of equations; however, they are complex and numerical models for integrating them stepwise in time and space use grossly simplified versions of the equations; require many hours running time; are still underdeveloped; and have not been adequately tested against data. However, they are conceptually sound and it is therefore important to design field experiments to test and to "calibrate" them. The basic variables are usually expressed as simplified versions of key measurements and are called "parameters." Because statistical methods are used in the calibration procedure, they are also referred to herein as covariates. This approach should enhance the sensitivity of the statistical test. But in addition, in an exploratory experiment it would provide a means of calibrating a model for different cloud types and terrain influences, and a basis for better insight into the seeding mechanism. To illustrate these points, some features of the stable orographic cloud will be discussed.

2 STABLE OROGRAPHIC CLOUD

The physical features of a stable orographic cloud are presented schematically in Figure 1. First, the air mass is stably stratified or neutral. Thin layers of convective overturning may be present, but there is no deep embedded cumuliform cloud. The base of the cloud is low over the upwind slope and somewhat higher over the downwind slope. The air rises and cools adiabatically above the upwind slope, and some of its water condenses in the form of minute cloud droplets, too small to fall to the ground. All cloud droplets would evaporate on passing into the downdraft region over the lee slope were it not for the fact that some are converted by natural and artificial iceforming nuclei to ice particles, which grow to sufficient size (and terminal velocity) to fall to the ground. Ordinarily, not all the liquid water is converted to ice; therefore, the precipita-

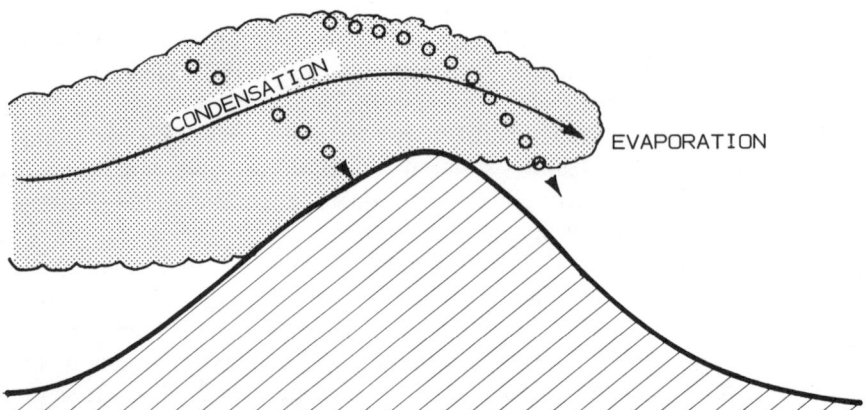

Figure 1 Physical features of a stable orographic cloud. The wriggly line is the cloud boundary. The solid line with an arrow represents mean airflow in the cloud layer relative to the mountain barrier. Dotted lines with arrows represent typical trajectories of ice particles falling through the cloud to the ground.

tion production mechanism operates at less than 100 percent efficiency. Since some of the water condensed on the upwind slopes is removed as precipitation, the condensation level is higher on the lee slope, hence, the higher cloud base there. If the process were 100% efficient there would be no cloud at all there.

The primary purpose of cloud seeding is to accelerate the conversion of cloud droplets to ice crystals by supplying artificial ice forming nuclei to supplement the natural ones which are usually in short supply. This procedure, when effective, captures more snow on the ground and leaves less cloud water to evaporate on the lee slope or in the downwind valley areas.

Detailed, systematic observation of all these features by means of cloud phyics aircraft, is a difficult, expensive, and hazardous undertaking because of severe turbulence, heavy icing and strong winds over the barrier. However, radiosondes released upwind of the barrier provide a reliable source of useful data. Air mass properties are adequately sampled even though the microphysical properties are not. The upwind cloud base and cloud top can be determined with considerable accuracy. The upwind flow pattern at all levels is accurately measured. By applying a simple orographic flow model, it is possible to infer modification of the air flow induced by the barrier, and the consequent rise in the cloud top. A network of ground-based sensors can provide a reliable source of information about precipitation, although this requires expensive maintenance under adverse weather conditions. Figure 2 shows various key parameters that can be estimated by means of the upwind sounding data.

The average precipitation over the barrier is roughly proportional to the mean cloud layer wind component normal to the barrier (VN). The greater this is, the more rapid the lift of the air, and the greater the rate of condensation. But the precipitation also depends directly on the moisture available in the air mass. This is best estimated from the difference between the saturated water content at cloud base (Q1) and that at the level of highest lift over the barrier (Q2). However, it is necessary to integrate up through the cloud depth (CD), and this becomes an important parameter. The product of (VN), (DQ), and (CD), where DQ = Q1 − Q2, is directly proportional to the total condensation rate over the barrier.

The efficiency with which the condensate is converted to ice crystals capable of growing and falling as snow to the ground depends directly on how cold the cloud top temperature is. This is because the concentration of natural (and also of artificial) nuclei increases exponentially with lowering cloud top temperature (CTT). Thus, the cloud top temperature is also a key parameter in the precipitation process. Parameters representing terrain configuration, such as the height of the barrier above the valley floor (H) and the width of the upwind slope of the barrier (L) are also of some importance, especially when transfer of results to another barrier is a consideration.

Using historical precipitation data, it is possible to relate by multiple regression techniques the barrier precipitation rate to these sounding-derived parameters. In the Colorado River Basin Pilot Project (CRBPP), carried out in the first half of the 1970s in the San Juan Mountains of Colorado, a very large amount of recording precipitation gauge data was collected and soundings were taken upwind of the barrier at intervals of every three hours during storms. Six hour precipitation blocks were related by regressions to the sounding-derived

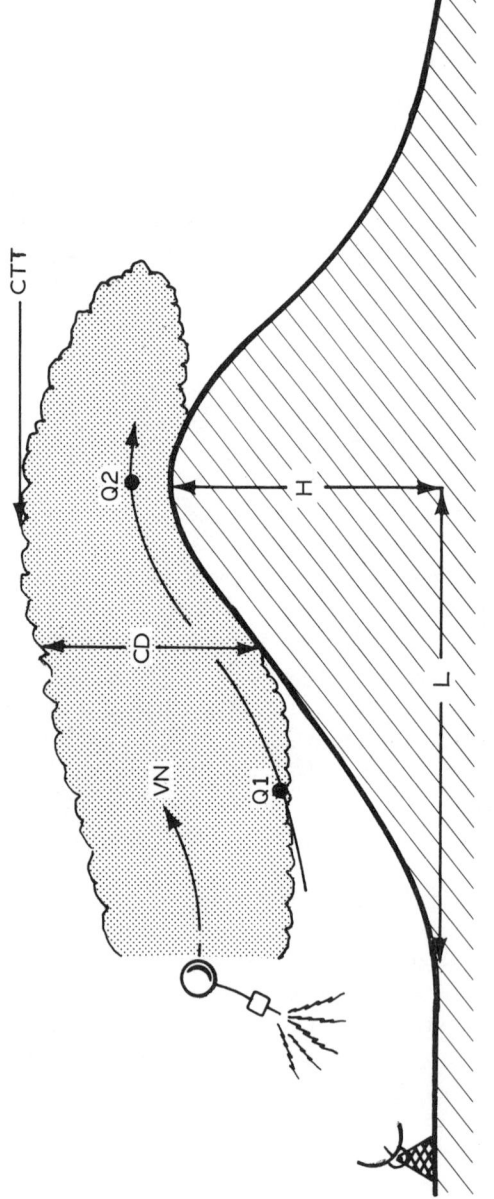

Figure 2 Sounding-derived and terrain parameters. The radiosonde released over the upwind valley is depicted by the balloon and ground receiver. The motion of the air from cloud base at Q_1 (saturated mixing ratio) to highest point of lifting over the barrier at Q_2 is shown by the long solid arrow. The short arrow represents the mean mid-cloud component of the wind normal to the barrier (VN). The mean cloud depth (CD), the height of the crest above, upwind valley floor (H), and the breadth of the upwind slope (L) are also indicated. The lowest cloud temperature is at the point indicated as CTT. The difference between Q_1 and Q_2, (DQ), represents the moisture released in an air parcel. VN, CD, DQ, and CTT are sounding derived parameters. H and L are terrain shape parameters.

parameters. Multiple regressions were formed on the parameter values and their squares. The multiple regression coefficients ranged from about 0.50 to 0.80 (Elliott et al., 1976). The regressions for this and for other barriers all indicate that the greater the normal wind component, the deeper the cloud, the greater the moisture holding capacity of the air mass, and the colder the cloud tops, the greater the precipitation rate.

When the clouds are seeded the barrier precipitation is increased due to added conversion to ice (glaciation) as mentioned above. But it can be decreased by another seeding-produced mechanism. With seeding the available water is shared over more crystals, and their sizes are therefore smaller, leading to lower terminal velocities and flatter crystal trajectories. More ice crystals blow beyond the crest before reaching ground level. This removes precipitation from the upwind slope, redistributing it to the lee slope. This is referred to as the "blowover effect." Under certain circumstances there can be a net loss over the barrier as a whole because many of the ice crystals blown over to the downwind slope evaporate in the dry subcloud layer before reaching the ground. This condition occurs in clouds that have very cold tops, and are naturally quite efficient. The cloud is then said to be "over-seeded." These various effects have been observed in the CRBPP (Elliott et al., 1978) and in several other experimental projects (Vardiman and Moore, 1977) and do conform to expectations based upon fundamentals. They can be assembled together to form a conceptual model, the basic features of which are illustrated in Figure 3. In the upper panel the solid line C shows how the condensation varies directly with normal wind speed (VN). The curve is for the case where the other parameters have average values. It also varies directly with cloud depth (CD), or water available (DQ), according to the middle panel. A direct relationship is shown against lowering cloud top temperature (−CTT) in the lower panel. The solid CG, on the other hand, represents the fraction of the condensate that is glaciated. In all panels this line falls beneath C, and this is the usual case, i.e., the precipitation process is less than 100% efficient. The solid line marked BCG represents a further adjustment to account for any blowover effects in natural clouds. For example, the stronger the wind, the greater the loss (top panel). With a deep cloud (middle panel) on the other hand, there is some increase due to greater growth of particles in deep clouds, and steeper trajectories. There is a large loss due to blowover effects with lower top temperatures (lower panel). This situation occurs in very cold top orographic clouds. The heavy solid line marked BCG represents the barrier mean precipitation rate.

The foregoing account pertains to natural clouds. We shall refer now to the dotted lines which represent the condition in seeded clouds. The CG line is seen to lie above that for the nonseeded case in each of the panels. This means that the fraction of the cloud droplets that are glaciated is always greater in the seeded cases. However, seeding does lengthen the particle trajectories so that there is a greater blowover loss. This shows up most markedly against the cloud top temperature parameter (lower panel) where under cold cloud top conditions the blowover losses exceed any gains due to the increased glaciation. This is the case of overseeding. There is an interaction also with VN, that is to say, with a greater VN the blowover loss due to seeding is amplified. There is

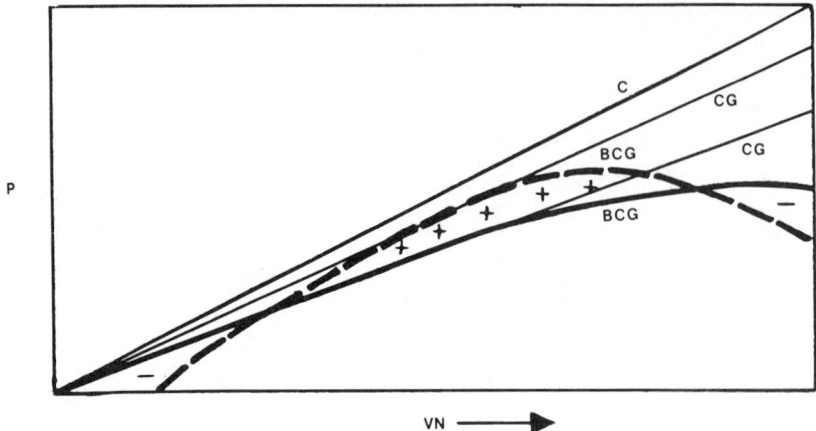

VN →

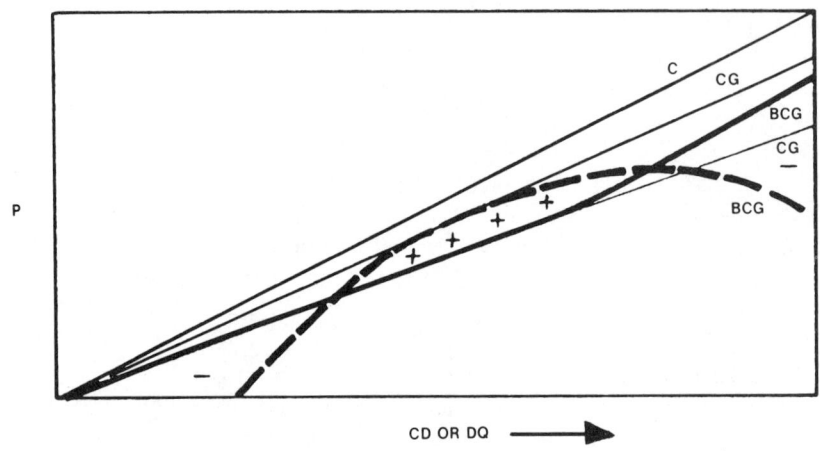

CD OR DQ →

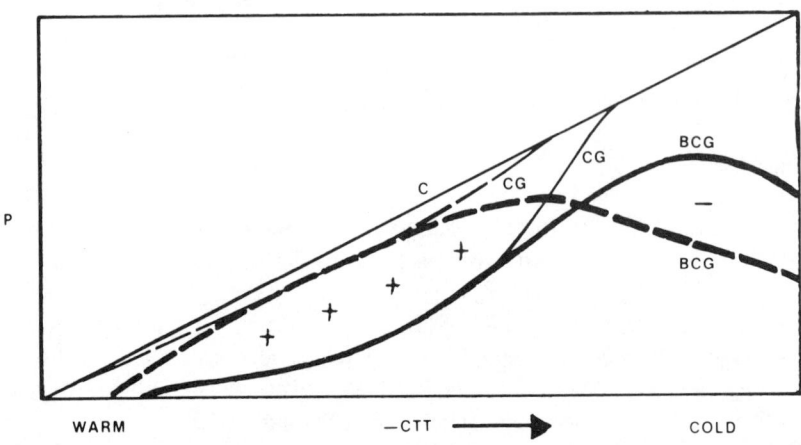

WARM — −CTT → COLD

also an interaction with the cloud depth, there being a greater blowover loss due to the greater fall distance from cloud top near the crest.

The difference between the seeded and not-seeded BCG curves represents the seeding effect spread out over the range of parameter values. The small pulses are in areas where seeding enhances precipitation and the negative ones are in areas where it decreases precipitation. It is seen that for high parameter values the blowover losses become more important than the gains due to enhanced glaciation. Overall the precipitation rate increases in a quasi-linear fashion with increase in the parameter values, but the seed/no-seed difference displays subtle second order effects. In multiparameter space there is an upper boundary to the positive effect area or "window." On the lower end of the scale, some negatives are also shown. Field experiments have clearly demonstrated that very shallow clouds can be overseeded using a normal dosage rate. With very light winds the flux of particles is reduced and concentrations enhanced so that overseeding may occur there also.

3 CUMULUS CLOUDS

The foregoing discussion is focused on the stable orographic cloud, a rather simple system. Much of the precipitation over mountains contains embedded cumuliform clouds, particularly over the western ranges closer to the Pacific Ocean. In much of the United States the majority of the precipitation falls from more or less isolated summer cumulus clouds. Figure 4 is a schematic representation of an isolated precipitating cumulus cloud. There is an updraft zone created by the air mass instability in the heart of the cloud where condensation is the most important process. On ascent to higher levels, where more natural nuclei become effective, there is a conversion of some of the cloud droplets to ice particles which grow and fall out as precipitation. Those failing to fall out are carried into the anvil along with any remaining unfrozen droplets. Even though these latter may completely freeze at the very cold temperatures in the anvil, almost all evaporate before they can fall to the ground.

The cumulus updraft plays a role similar to that of the normal wind component in the orographic case. In the cumulus case, the updraft is related to the potential energy available in the air mass and this can be determined from the sounding plotted on a thermodynamic diagram. Seeding effects are also related to cloud top height and temperature. Project Whitetop (Flueck, 1971)

Figure 3 (Opposite) Components of the barrier orographic precipitation rate. Each of the 3 panels contains a set of curves representing various components of the mean orographic precipitation rate (P) on the upwind slope of a barrier. The top panel shows how the components vary against increasing normal wind speed (VN). The middle panel shows the variations against cloud depth (CD), or moisture available (DQ). The latter two are usually well correlated and therefore are about the same. The lower panel shows the variation against the cloud top temperature, with colder tops lying to the right. The components are the barrier total condensation rate (C), the total condensation rate that is glaciated (CG), the total condensation glaciated, corrected for blowover losses (BCG). BCG is the barrier precipitation rate: The components are shown for the natural case by solid lines, and for the seeded case by dashed lines.

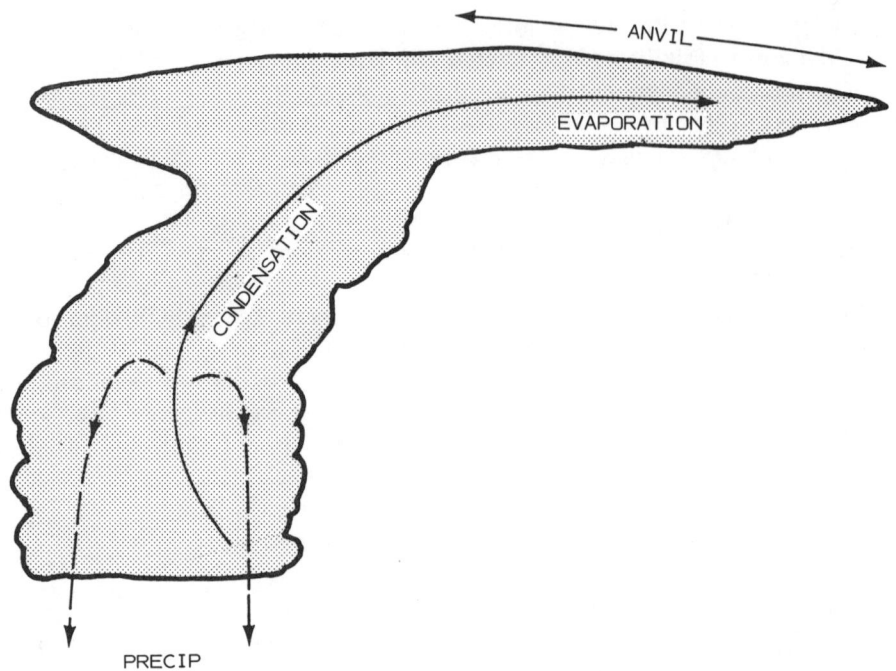

Figure 4 Physical features of a cumulus cloud. The main updraft is indicated by the solid line with arrows. At highest levels this current moves laterally into the anvil. Precipitation particle motion is indicated by the dashed lines and arrows.

results have indicated this to be the case. When the radar cloud tops were high there were decreases in precipitation; whereas, with the warmer or less high tops, increases were indicated.

Isolated cumulus clouds react to seeding in more complex ways than do orographic clouds because of dynamic effects. The release of heat of sublimation within a convective updraft adds to the buoyancy of the updraft and therefore to its rise. This means that not only can one hasten the glaciation within cumulus clouds, but one can also increase the total of condensed water. Usually this, in turn, enhances total precipitation, but if the updraft is increased too much, the flux of the converted ice particles into the anvil is increased and something similar to the seeding produced blowover loss of the orographic case occurs. These remarks also pertain to cumulus clouds embedded within an orographic cloud. In the San Juan there appeared to be a tendency to redistribute precipitation in this way over to the lee slope, with a net reduction in barrier total precipitation. In the Sierra Nevada mountains, with their broad upslope region, the same redistribution effect would lead to a net increase in barrier precipitation.

An even more profound dynamic effect is often observed where neighboring clouds merge with the seeded cloud, and thus develop a larger mesoscale circulation with a greatly enhanced low level moisture convergence, and a resultant increase in precipitation rate, area, and duration. This effect was noted in the Florida Area Cumulus Experiment (FACE) (Woodley et al., 1976). The seeding of convection bands in the SBA II experiment appeared to increase their width, and therefore, the precipitation duration at a point (Elliott and Brown, 1971). In order to evaluate such effects in an exploratory experiment, it is necessary to analyze continuous records of ground level precipitation, and continuous records of radar return. Aerial observations can also provide information, particularly of any changes in cloud size and updraft strength.

In the orographic situation, embedded convection may be organized into bands which have their own mesoscale circulation. This is the case in the ongoing Sierra Cooperative Pilot Project (SCPP), and was the case in SBA II. Seeding treatment of convection bands can alter the barrier water balance because of dynamic as well as microphysical effects.

The foregoing has been offered to illustrate the potential value of using carefully formulated covariates; not just to sensitize the evaluation of results; but also as a means for interpreting results for various cloud types in different terrain settings. It has been assumed that the latter feature would be essential to an exploratory experiment. However, it could also be piggy-backed onto a purely confirmatory experiment designed simply to test whether precipitation had been enhanced by the best available treatment technique. The key is having suitable model plus suitable observations available from which to fashion meaningful covariates.

BIBLIOGRAPHY

Elliott, R. D. and Brown, K. J. (1971). The Santa Barbara II Project—Downwind Effects—*Preprint Volume—International Conference on Weather Modification, Canberra, Australia, Sept. 6-10, 1971.* American Meteor. Soc., Boston, Mass., 179-184.

Elliott, R. D., Shaffer, R. W., Court, A., Hannaford, J. F. (1978). Randomized cloud seeding in the San Juan Mountains, Colorado, *J. Appl. Meteor.*, 17, 1298-1318.

Elliott, R. D., Shaffer, R. W., Court, A., Hannaford, J. F. (1976). Comprehensive Evaluation Report, Five Winter Seasons, 1970-1971 . . . 1974-1975. Aerometric Research Inc. report to U.S. Bureau of Reclamation, Dept. of Interior, Denver, Colorado.

Flueck, John A. (1971). Final report of Project Whitetop: Part V—Statistical analyses of the ground level precipitation data, Univ. of Chicago Dept. of Geophysical Sciences, 294 pp (available NTIS N72-13559).

Vardiman, L. and Moore, J. (1977). Generalized criteria for seeding winter orographic clouds, Skywater Monograph No. 1, U.S. Dept. of Interior, Bur. of Reclamation, Div. of Atmospheric Water Resources Management, Denver, Colorado.

Woodley, W. L., Simpson, J., Biondini, R., Sambataro, G. (1976). On NOAA's Florida Area Cumulus Experiment (FACE) Main Rainfall Results, 1970-1975. Preprint Volume—*World Meteorological Organization—Second WMO Scientific Conference on Weather Modification*, Boulder, Colorado, 2-6 Aug 1976, 151-158.

Some Statistical Aspects of Weather Modification Studies

Oscar Kempthorne

Statistical Laboratory
Iowa State University
Ames, Iowa

Abstract

After introduction and initial discussion of the general problem, it is argued that there has been partial failure by the cloud physicists and by the statisticians to bring to bear essential ideas. General ideas for fertile comparative experiments are discussed, in particular, the concept of experimental unit and its use, delivery of the stimulus and measurement of yield. The role of randomization with respect to judging whether there are treatment effects is discussed. Comments are given on the role of multivariate data analysis, data searching and crossover designs. The essay closes with comments on the Florida State statistical effort, and on the presentations that were given at the conference.

1 INTRODUCTION

Because this essay is directed both to atmospheric scientists and to statisticians, it is appropriate that I give a little background, some of which is personal. From around the end of World War II, when Schaefer and Vonnegut discovered that the introduction of "seeding" material into a cloud caused a remarkable physical change that cannot be doubted, there have been continuing attempts to use this discovered process to control weather, e.g., to produce precipitation, to reverse the growth of hurricanes, and so on. Strong claims have been made that precipitation can be induced, and equally strong (or weak, if I may say so) attempts to confirm the claims have in general failed to produce the desired confirmation. In the late 1950s the National Science Foundation set up an advisory panel on weather modification, and I served on this panel for a few years. At that time, I did my homework to the best of my ability. I became acquainted, more or less, with the history of the efforts to understand rain-making. Then I became involved in other pursuits. I was not drawn back to the field again until the winter of 1978. My first return consisted of reading the bibliography produced at Florida State University (Hanson et al., 1976). I have no idea of the amount of public monies that have gone into the area in the intervening twenty or so years. But it is surely very appreciable. My general reaction was one of distress that so little had been accomplished. I attribute this lack of progress to three origins: (1) the immense difficulty of the area, (2) partial failure of the physical scientists involved, and (3) partial failure of the statisticians involved. I now give brief amplification of these views. The body of the essay will deal predominantly with the third.

2 THE DIFFICULTY OF THE PROBLEM

This will become clear in the body of this essay. However, it is obvious that a cloud is an heterogeneous and ill-defined entity and a weather system passing over a geographical region in a particular time period is even more heterogeneous and ill-defined. There are obviously huge temperature gradients vertically; there is obviously huge physical variability with regard to density of water in some form both horizontally and vertically. This description of difficulty could be extended nearly indefinitely by the atmospheric expert. So that there has been rather small progress should be no surprise. This is not a "simple" problem like feeding a farm crop or an animal or a human. And I put the word simple in quotation marks because these problems are not in fact simple. We need only look over the efforts of the past 150 years with innumerable scientific studies and comparative experiments, and we see that huge progress has been made, but also we see that investigative effort has uncovered scientific and applicational questions by the "ton." Science never reaches *the* answer. So that there has been little progress is not to be marvelled at; rather we should take a little satisfaction from some improved scientific understanding having taken place.

2.1 The Partial Failure of the Physical Sciences

In addressing this, I do not have in mind failure with respect to the physics of rain formation for instance. Indeed, I am not competent to judge this at all. What I have in mind, rather, is that there has been, I judge (and perhaps erroneously), a failure to appreciate the complexity of the situation that I refer to cursorily in the previous section. Everyone knew long ago about the heterogeneity of, say, a cloud. Everyone knew that nucleation depends, in a very obscure and not yet well understood way, on the fine structure of the cloud (using the term in a general sense). Everyone knew that the motion and identity of a cloud is little understood.

What was the nature of early attempts at weather modification? One took what seemed to be an identifiable portion of the atmosphere. One placed ground burners at some points; one then lit these, and a huge number of, say, AgI particles poured out. They went upward at first to be sure. But where did they go after that? In the early days it seemed, and I judge to a considerable extent it still seems, that one need not worry about such questions. After all, some huge number of particles was emitted. So there could not be any relevant and important question to what happened to them. In my recent rather cursory review I note that by the middle of the 1960s this question was addressed, but I incline to the view that we still know incredibly little. The questions are, of course, not easy to attack.

If we pass over such purely physical questions, we must surely face the fact that even if we can speak about a cloud as an entity with some permanence over time and sufficient permanence to regard it as an experimental unit that is given a stimulus, there is huge variability among a set of clouds that seem on the basis of our observations to be very much alike. This sort of problem has been faced in biological and agricultural science for decades. Can one find two pigs, or two atmospheric scientists, or two plots of land that are sufficiently alike that one can treat them for experimental and comparative purposes as being so nearly alike that we can regard them as identical? Of course not. Even the not-so-intelligent-man-in-the-street knows this. What has been the universally used process to attempt to acquire knowledge of the effects of stimuli in this situation? The neophyte of biological or agricultural research knows that there is only one data collection process that can enable *partial* progress in the face of this unavoidable lack of identical units and it is randomization. This has been known for some 50 years. Yet the use of randomization was introduced into the weather modification area only in the late 1950s; if we may make a judgment from the report of the Statistical Task Force (1978), the argument still has to be presented. It is done there with a whole new jargon as though atmospheric scientists must be given ideas that are very old with new terminology. Perhaps this is necessary for the psyche of the field. I do not know. But the reaction of a trained experimenter in the "noisy" sciences (and I do not mean noisy only in the sense of measurement or quantification) is that what is given should be standard knowledge of the neophyte. It is profound disrespect to the overall process of science not to give cognizance to the mentation of the past. The consequences of developing different terminologies for a process in different substantive fields are obviously deleterious at best.

2.2 The Partial Failure of the Statisticians

Anyone who has been a real practicing statistician knows that the critical part of the practice of statistics does not lie in the well-formed questions of mathematical statistics. An example may be useful. A well-formed question in mathematical statistics is the following: Suppose I have a random sample from $N(\mu, \sigma^2)$. What, then, is the distribution of $x_U - x_L$, where x_U is the largest and x_L is the smallest x in the sample? The "mathematical" statistician is trained to be able to attack this rather simple problem and, of course, problems of much deeper complexity.

The statistician who wishes to *apply* ideas of mathematical statistics will automatically raise questions that are prior, in the whole scientific process, to the well-formed question of mathematical statistics. What are these questions? Every potential substantive application raised a distinctive set of questions. I will give some that come to my mind in "analysis" of a weather modification comparative study:

(1) What was the experimental unit?
(2) What were the treatments?
(3) How did you assign treatments to experimental units?
(4) How do you *know* that the treatments were delivered to the units?
(5) How did you measure the response of a treated unit?
(6) What are the measurement errors of your responses?
(7) Can you regard the units you used as a *random* sample from a *definable* population?
(8) If so, what is that population?
(9) If you can, in fact define a population of possible units, what assurance do you have that you have a random sample? To bring home to you the force of this question, here is a collection of numbers. Can you, the scientist, envisage any way of judging that this is a random sample from a population? And to force the question to its ultimate depth, before you look at the numbers, which are on small cards, I will place the cards in an urn, shake the urn and then draw cards one at a time. You, then, will have the numbers in a sequence and you are to tell me how you make a judgment of whether the collection of numbers is a random sample from *some* population. The answer is, of course, that no one on this earth can answer the question. To continue this example is, I think, instructive. Suppose we take the collection of numbers, and obtain a histogram or a cumulative frequency plot, can you tell me whether this is a random sample from, say, $N(5,10^2)$? The answer then is less obscure, because it amounts to the following. *Given that you can assure me that you have a random sample from some population, I can aid your questioning of whether that population could be $N(5,10^2)$.*
(10) The upshot of the trivia of the previous paragraph is that we cannot determine from a set of numbers alone whether we have a random sample from some definable population.
(11) We may, however, be able to sustain an assumption of random sampling, if our experimental units arise from a dynamic process that we have observed, and we have analyzed data in all the ways that occur to us and cannot invalidate the assumption. We may then be able to take the position: On the basis of the history of the process I have observed, I take it to be reasonable to assume that I have a random collection of units.
(12) However, for weather modification experiments I have seen (and, obviously, I have not looked at every one, or even any one deeply), I suggest strongly that the assumption that Nature presents the weather experimenter with a random sample of units from a population of units is untenable.
(13) One consequence of this *for me* is, perhaps, worth mentioning. Neyman, in many places too numerous to cite, has advocated $C(\alpha)$ tests as being uniquely appropriate for the inferential problems of weather experiment. It follows from what I say earlier that I believe that this claim cannot be sustained. I wish to suggest that the search for "optimal tests" assuming random sampling from some mathematically

specified class of distributions is not the whole of the best solution to causal inference in the weather modification experiment. I do grant that this work may well lead to a good candidate test criterion for use in such an experiment.

(14) A consequence of this *for me* is that the appropriate way to make a test of significance of whether an apparent effect could have arisen by chance (the chance being that brought in by the randomized design) is by means of the randomization distribution that is generated by the randomization used in the choice of the actual plan of the assignment of treatments to experimental units. Gabriel (1967, p. 98) says "Only non-parametric methods based on randomization in the design will give exact significance tests," and used such methods. In my book (Kempthorne, 1952), in a paper (Kempthorne, 1955), and in various other papers through the fifties and sixties, this viewpoint was exposited. I am not, then, at all surprised by views about randomization and "rerandomizing" (= randomization testing) of the Statistical Task Force, which seem to be presented as novel powerful statistical contributions to weather modification statistics. I suggest, however, that that report is notably deficient with regard to my points (1) to (13) above. And this is one of the aspects that leads me to talk about the failure of the statisticians.

3 THE COMPARATIVE EXPERIMENT

Weather modification experiments are no different in logical (or illogical, if you please) nature from comparative experiments in biology, agriculture, medicine, or whatever. It is relevant, then, to evaluate weather modification experiments in terms of what I insist are standard ideas in all these areas. I think, in fact, that it is of critical importance to do so, because I believe the general basic ideas have not been generally appreciated.

To think about seeding clouds, I find it useful to make a wide analogy. I think of a cloud or a cloud system as being a pig flying in the sky. The problem is to determine if injection of substance A (e.g., silver iodide crystals) will produce a response. We shall regard the pig (or in other cases, a pen of pigs) as the experimental unit.

In the case of beef animals, it was suggested from some source that DES would improve the growth of feed cattle. The questions then arose of how one should deliver a DES stimulus and, of course, how strong a one. Then, with a method of applying the stimulus, one should have some means of verifying that the stimulus dose actually entered the relevant portions of the animal. Finally, one has to have a decent measure of response. The example is interesting because it turned out that a dose could be too large or too small. Delivery might be by injection and then if there is no variability in our "shots" and if we do the injection properly, we can be confident that we have delivered the treatment. It might be that delivery would be via food; and if our stuff is mixed well with the food then we can be confident that treatment was delivered to the alimentary canal of the pig. There might, of course, be problems. The stimulus might become locked to an inert constituent of food that merely passes through. But such a happening would be part of the outcome that is being measured and could explain a negative outcome. It might be that one should make an implant.

After some experimental period, we observe the "pig" in some way, e.g., weight, backfat thickness, and so on. One of my points is that the pig is obviously present in unambiguous form and nature throughout the experiment. We shall have, of course, many questions in our mind about what to measure and we shall have to validate our measurement process, obtaining a reasonable idea of the magnitude of measurement error.

In the above, I have used the term "experimental unit." The concept is rarely explicated, even though it is intuitively obvious. The idea is that a portion of space-time that retains identity over the duration of the experiment is given a particular treatment. The whole of the experimental material is partitioned into "pieces," and if a piece is such that the whole of it can receive one of the treatments in the experimental plan, then any such piece is called an experimental unit. The critical point is that in the experimental plan, which consists of allocating treatments to pieces, parts of a piece are not to be regarded as experimental units. To give an agronomic example, the plan may call for a certain piece of land receiving one or other of the treatments. Why, then, should one not regard, say the 4 quarters of this piece as being experimental units? This can be answered at two levels. If one uses a model: observation = true value + error, then it is obvious that the quarter pieces have commonality of error. At another level, if one adopts the viewpoint that one should use a randomized design and that one should use the distribution of results that would arise, conceptually of course, if one repeated the experiment to evaluate the actual result one gets, then the quarter pieces would "tag along together" in this randomization distribution. To treat the quarter pieces as experimental units and to use the variability among them as a measure of error, would be wrong as Fisher (1935) exposited, because in comparing treatments we are comparing pieces, and we have to obtain a measure of variability between pieces. The variability of quarter pieces within pieces has no bearing *at all* on the variability between pieces.

A further critical ingredient of this prosaic pig, cattle, or field plot experiment is that there must, if the analysis is to follow conventional lines, be no interaction of experimental units. No unit can influence a neighbor unit. So in these "old" experimental areas, there was extensive work on examining optimum size of pen or of plot, within field plots, for instance, the use of guard rows.

Then, of course, there was extensive work on the yield measurement process. One might, for instance, use sampling of experimental units; so studies were made of modes of sampling to obtain reliable and sufficiently accurate schemes.

I now address the "typical" weather modification experiment in terms of these ideas.

3.1 What is the Experimental Unit?

I do not have knowledge of the whole span of the experiments. St. Amand at the Statistical Workshop at Florida State University in October 1978 described in an oral report non-designed intervention activities with particular clouds, and was able to observe what happened to the cloud, e.g., how it formed an anvil, or, perhaps, broke up into two parts. It was obvious in these, incidentally, that the intervention produced dramatic effects, though I suppose that the total skeptic would want a randomized experiment, because clouds may form anvils or break into two parts without intervention. I wish, indeed, that such experiments would be done on small, definite clouds or "pigs" flying in the sky.

The so-called experimental unit has often been a very vaguely defined (if at all defined) region of the atmosphere above or downwind, or some such, from a stimulus delivery apparatus. Does this experimental unit hold its identity until a response can be measured? The question is obviously rhetorical. It does not. How about convective bands in a Santa Barbara II study? Elliott informed us at the Florida State workshop that he has the opinion, fairly strong, that these convective bands do retain identity over the experimental period. I believe that a main role of a statistician is to be skeptical, to "push the scientist to the wall," and I find that I am rather skeptical. But in saying this I must emphasize that I have had no opportunity for the sort of aggressive and unpleasing questioning that is necessary, and is, I believe, essential.

Is the experimental unit defined in the bulk of so-called weather experiments? I believe the answer must be "no" and assertively. The experimental unit is, at best, a combination of a time point or brief interval with a fuzzy "pig" flying in the sky. The "pig" is not at all well-defined as regards boundaries. It has a very obscure and uncertain life over time. It is here as a definite, perhaps, region of space at time t_0; at time $t_0 + x$, it is somewhere else in space; we may not even be able to identify it as being the after-cursor (c.f., pre-cursor) of the cloud we *think* we have seeded. How can we possibly claim to have an analog of our real pig experiment or our real rat nutritional experiment or whatever?

Are there other sorts of experimental units? A somewhat common one can be characterized as the "seeding or non-seeding day." We take a geographical area; one part of this we define to be the seeding area and another part as the measurement area. We seed or not seed on a particular day in the former, and measure yield, precipitation, in the latter. We then do data analysis without or with additional measurement data such as yield in a so-called control area. Or we might have two areas, one of which is seeded and the other of which is used as a yield or response areas. Is this a viable concept of experimental unit? I am very skeptical. It is, perhaps, all too easy to be critical of what has been done in the past. So I must emphasize that I do not wish to cast aspersions. The way to knowledge is tortuous and "boot-strap." We do what we can, and in every field of science we can look back and carp, saying to ourselves, "How silly and stupid!" To do this is to exhibit an essential total lack of understanding of scientific method. The idea of the present paragraph is interesting. It had to be tried, I suggest. It is certainly true that one can conceptualize the situation so that the whole region for a day (though there is obscurity on this, because we can ask if it should be 24 hours or 8 hours or whatever) is the experimental unit. The resulting data then have the logical form of a randomized comparative experiment, and, I suggest, an entirely proper mode of forming a judgment of whether yield has been influenced by seeding is to perform a randomization test of significance, as was indeed, done in Israel (Gabriel and Feder, 1969). This paper exposits the whole of the approach.

My overall reaction to this, then, is that it had to be tried. The effects of seeding, by the admittedly naive method of delivery of the stimulus might have been large enough and consistent enough that good evidence for effect of seeding would have been obtained. I am *not at all* critical of this effort. I regret that the phenomenon is not so simple and repeatable that this type of effort was not clearly successful.

It is appropriate, I believe, at this point to present a view. One starts, shall I say, from a physical science viewpoint and does a comparative experiment; i.e., some units receive seeding and some do not. After several trials of this type, one realizes that there is huge variability among the units. One then hears about randomization and becomes convinced that it has merits. The next step is then all too obvious. One introduces some sort of randomization. Then one says "You cannot criticize my experiment because I randomized," as though the mere fact that one has randomized makes the whole affair valid and highly scientific. As I have exposited, perhaps ad nauseam, in too many places to cite, randomization is not a panacea. If it does anything, and I believe it does, it controls statistically extraneous forces that have been randomized over, and no more than this or no less than this. If these views, are to be given any weight, one must, then, respond to the scientist who says proudly "I randomized" by asking "What did you randomize over?" and by stating that randomization is only one, very critical to be sure, component of a "good" experiment when there is variability among units.

3.2 What About Delivery of the Stimulus?

There are problems, obviously, if only because our experimental unit if so imprecisely defined. But, event if we let this pass, how about delivery of the stimulus? The early experiments were a clear case of extreme naivete — natural enough with the difficulties, but still incredibly naive. What was the procedure? Set up a line of burners and set them going. The burners spew out some huge number of AgI particles. Where do the particles go? It would seem that nobody knows. Do they get to the vaguely-defined cloud system? Who knows? As far as I can see, the experimenters do *not* know. I was interested that Court (1967) said "The actual intensity of treatment cannot be specified even as to order of magnitude." Also, I quote Simpson (1974): "Regarding targeting, many cumulus experiments have been conducted by distant releases, without proof of what concentrations, if any, reached the intended active portions of the clouds." How about delivery from planes? What happens? Who knows?

3.3 What About the Measurement of Yield?

The difficulties are obvious. The vaguely defined original experimental unit has become even more vague. God knows, perhaps, where the spewed out silver iodide crystals went. Man surely does not. What is the yield of the unit? Who knows? We have a miscellaneous collection of rain-gauges, not set up to reflect the rainfall of a *region*. We cannot have anything but the very vaguest idea of what yield is.

4 ARE THE ABOVE IDEAS APPRECIATED IN THE FIELD?

I have not read the literature exhaustively. A reaction may be that I am merely saying the obvious.

What was the experimental unit in past studies, e.g., that of Santa Barbara I? (Neyman et al., 1960). I do not have the original account at hand. However, it seems that

(a) there was some defined (or vaguely defined) region in space near Santa Barbara;
(b) 12 hour units of time were considered;
(c) if a configuration of clouds at some time, perhaps 8:00 a.m. of a morning, was judged to fall in a class of circumstances, then the cloud system over some region (what region?) was judged to be an acceptable experimental unit;
(d) by randomizing, e.g., a coin flip, it was decided whether to seed or not to seed.

The method of seeding was "ground generators." The response variables were amounts of rain given in rain gauges (somewhere definite of course).

Then, presumably, some average of some rain gauge yields was taken to be the yield variable.

And finally, an F test was used.

I do not want to be critical of Neyman, Scott, et al., but I suggest strongly that if one thinks about

(a) definition of experimental unit,
(b) imposing the stimulus and the question of where the stimulus went,
(c) permanence of the experimental unit,
(d) measurement of yield,

then it is not at all surprising that the noise level in the experiment was great.

I imagine that this was the *only* sort of experiment that was possible, so no blame should be attached.

If there were, in fact, a huge effect of seeding, more or less constant on some scale (e.g., logarithm yield), the experiment may well have demonstrated the existence of effect of seeding. So as a pilot experiment it should not be criticized.

Quite apart from showing that there is a seeding effect, this experiment is surely valuable in showing the noise level that will be encountered in an experiment of this kind.

But repetition of experiments of this kind is surely a waste of effort.

5 EXPERIMENTAL DESIGNS

In statistical teaching to hundreds of biological and agricultural scientists at my own institution (and this is obviously only one of many examples), the ideas of completely randomized designs, randomized block designs, Latin square designs and so on are routinely taught. I find it ironic, then, that these common ideas are given new names by the Statistical Task Force. We have, in fact, something "new" designated as OJR blocking. Indeed, I am appalled. It has been *totally commonplace for decades* to use subjective and objective criteria for placing experimental units in blocks, e.g., of two if one had twin pairs of animals, and then randomizing the assignment of treatments. Then one would, or at least I would, use a randomization test as a partial basis for forming an opinion as to whether the treatment difference has produced a response difference. Very often, I would not go to this trouble, relying on a normal law approximation, as exposited, (Kempthorne, 1952, Chapters 7 and 8, and particularly Section 8.4). I have to confess to some wonderment on whether people do a reasonable amount of homework, hard and time-consuming, though that is.

These are, however, additional aspects that were hinted at in my "ancient" presentation. There is the big problem of interaction of treatments and experimental units (loc, cit., Section 8.3). I now am of the opinion that even if one has a validated basis for blocking, one should always consider the potential utility of what may reasonably be termed "super-blocks" (though I do not have the arrogance to suggest that this term should be made part of the professional language). In these the block size could be with 2 treatments, for instance, 3 or 4 (this case and generalizations thereof having been discussed by Wilk (1955).) The idea is that the ordinary complete block design results, with the ordinary analysis, in a complete confounding of variability between units and interaction of blocks and treatments. I and former students have written, again perhaps ad nauseam, on this, and I shall not cite references.

It is obvious, I suggest, that *right from the beginning* the possibility of treatment-unit interactions should have been entertained.

5.1 How Much Randomization?

It is clear, from the simple logic of randomization and randomization tests, that one could merely construct a list of possible experimental plans, eliminating ones from a complete randomization that seem "offensive" or "dangerous." One would then go through the remaining steps as usual with the given limited set. This is in fact what is done, to some extent, by using a block experiment rather than a completely randomized design.

One must, however, note that if one thinks of a conventional analysis of variance, then one will be subject to the problems of estimating standard errors that Fisher exposited so well, vis-a-vis systematic designs, in *The Design of Experiments*. If one is successful one will have reduced standard errors, but an ordinary computation will show increased standard errors. The situation with using constrained or restricted randomization which was written on by the late W. J. Youden is not, for me, as clear as I like. [A reference is Sutter et al. (1963), but it only broaches the topic.] An aim must also be to get a point estimate of effect and a standard error, and this may present difficulties with an incomplete randomization set.

5.2 Blocking and Covariance

A discussion has been carried on in the literature for some five decades about the role of blocking and covariance. I shall not attempt to review this for its relevance to weather experiments. There are no simple answers in this area. However, it is elementary that blocking, which some might wish to call stratification, of experimental units has two roles. The one commonly mentioned is to reduce variability among experimental units *within* the groups within which treatments are assigned and then treatment comparisons are made. Another role is that blocks should, in a new area of investigation, contain some repetitions of treatments. I find it ironic that Fisher in his classic book *The Design of Experiments* enunciated the idea that we should obtain error for treatment comparisons from the variability of units (plots) treated alike, but *in fact* Fisher in the whole book did not give one design with which this is possible. In the ordinary randomized block design, variability "between units treated alike" (except, and critically, they are in different blocks) is totally confounded with block-treatment interactions. This tells us that the second role of blocking is to determine if there is any evidence of treatment interaction with blocks, this being a classification on some rational, even if subjective, basis of the units. So one might wonder, for instance, if seeding has differential effects with two different types of cloud.

A supposed alternative to such blocking is, of course, the inclusion in the model of analysis terms that take account of potential differences in types of unit. So one might include one or more categorical covariates or arithmetic covariates. The usual procedure, then, is to "charge ahead" with normal probability distribution theory. I have had considerable doubts about whether such normal theory mirrors or approximates randomization analysis. I am sure one should have doubts. However, I now take the viewpoint that if one is doing such covariance analysis, one should go through the full randomization test procedure for the analysis of covariance criterion that one uses. I do not feel hesitant about this now because of the advent of high speed computers, and because if we do have a big randomization set, we can fairly easily draw off a random 200 or 1000 members of that set, as was done by Cox and Kempthorne (1962).

In this use of covariance, it is clear that the covariates give information on the experimental units available *before* treatments were imposed.

The search for covariates for this use must obviously be based on theory or on data analysis of previous studies that are judged to be similar.

5.3 Multivariate Analysis of Data

The sort of covariance analysis I indicate in the previous section is, of course, a form of multivariate analysis *with the knowledge* that the covariate observations could not have been affected by treatments.

The situation is entirely different when we have a multivariate response measurement. The question of whether the supposed covariate has been influenced by treatment is obviously critical. One should, I suggest, examine this question by randomization analysis of the covariate observations in relation to the treatment allocation design.

I judge from my limited reading of weather literature that the status of variables on whether they can be treated as covariates in the design sense is often very doubtful.

This topic raises another related question that is not easy to resolve. Suppose we have a multivariate response, then whether or not certain components can be regarded as being unaffected by treatments, it is still relevant to ask what function or functions of the whole "response" variable should be examined. I suggest that unfortunate remarks have been made on this, e.g., by Fisher in his classical text. If one should use some "decent" physical theory to construct a yield variable, then one would be using, in Fisher's terms, an "arbitrary" correction, with the obvious implication that no decent scientist would be "caught dead" using an arbitrary correction. I need only say that I reject this view. One should clearly use all the scientific knowledge one possesses.

5.4 Post-Stratification or "Post-Blocking"

I need to say essentially nothing about this, because it is covered in the previous section. In doing this, one is adjoining additional "model terms" to one's tentative explanation. Because the aim of the study will be to assess evidence for treatment effects, one should again use randomization tests related to the treatment allocation plan.

5.5 Data Searching

A perennial problem in the analysis of any planned experiment is the utility and validity of searching the experimental data, and, with the idea (a good one, I judge) of trying to establish significance, one then digs and digs. The inevitable outcome, of course, with the data being finite, is that if one searches long enough one will find "significant at the 5% level" effects.

My view is that there are two classes of experiment (not of data analysis, though part of the analogy may be correct), namely, *exploratory experiments* and *confirmatory experiments*. I take it to be obvious that, if one uses words properly, one must set down a priori and follow totally without deviation a protocol of experimentation and data analysis in the confirmatory experiment.

The point about significance tests in a confirmatory experiment is that they are "acts of intention" (to use a phrase that a colleague I.M.Z. van Aarde likes). To determine what significance to attach to, say, a P-value, one must ask, "Why did you make that test of significance?" If the answer is (a) Because I planned to or (b) Because something totally outside the study prompted me, then the P-value one gets has evidential significance as it stands.

If, however, one has been led by the data to make a certain test of significance, the evidential import of the result is very questionable. It is elementary that the way to "water down" the significance of the apparently most significant of k independent tests is to take

$$P^* = 1 - (1 - P)^k$$

Simple calculations, after 10 searches leading to a criterion with P-value equal to 0.05, tell us that we should, perhaps, really attach a significance level of $1 - (0.95)^{10}$ which is about 0.4.

So if massive search leads to a criterion with a nominal P-value of 0.05, the only appropriate reaction is to shrug one's shoulders, if the intention is confirmation.

It is obvious, of course, that one should not use results of an experiment only for examining prespecified questions. The whole scientific "game" is to do many analyses to obtain ideas which can be subjected to confirmatory tests in new studies. It is interesting to speculate how one can, in general, "water down" apparent degrees of significance. I know of no good recipes.

5.6 Crossover Designs

A strongly advocated design for some sorts of rain making experiments is the so-called crossover design. With the emphasis that I grant, and apparently, others also grant, to randomization inference, it is appropriate and potentially useful to discuss briefly randomization analysis of such designs. Beginnings of strict randomization approach to this was given by Kempthorne (1952, Chapter 29, entitled "Treatments Applied in Sequence"). Some elementary description of the ideation may serve some useful purpose.

I judge that the only strict randomization analysis of this goes as follows. I shall consider a completely randomized design with two "treatments."

I have $N = 2m$ experimental units. Each unit consists of an ordered pair of 2 sub-units. We can then index units by $i = 1, 2, \ldots, N$, and then sub-units by (i,j) $i = 1, 2, \ldots, N$ and $j = 1, 2$. We have two "treatments" that we index by $k = 1, 2$, and each "treatment" consists of a pair of treatments: treatment 1 consists of applying stimulus 1 to, say, the first of the sub-units with stimulus 2 to the other sub-unit, and treatment 2 consists of the reverse.

To proceed, I suppose that sub-unit (ij) with stimulus k gives a number y_{ijk}. One can adjoin a random measurement error, this not being involved in the randomization analysis except as additional, hopefully uncorrelated, random error.

Now let

$$\delta_i = 1 \text{ if sub-unit (i1) receives stimulus 1}$$
$$= 0 \text{ if sub-unit (i1) receives stimulus 2}$$

with the consequence that if $\delta_i = 1$, then sub-unit (i2) receives stimulus 2, and if $\delta_i = 0$, sub-unit (i2) receives stimulus 1.

The population of numbers $\{y_{ijk}\}$ is the population that is being sampled, with this special mode of sampling.

Now let z_{i1} = observation on unit i with stimulus 1, and z_{i2} = observation on unit i with stimulus 2.

Then it is obvious that

$$z_{i1} = \delta_i y_{i11} + (1 - \delta_i) y_{i21}$$
$$z_{i2} = (1 - \delta_i) y_{i12} + \delta_i y_{i22}.$$

Our statistical operations will be performed on the set of numbers $\{z_{ik}, i = 1, 2, \ldots, N; k = 1, 2\}$.

What statistics shall we calculate? Obvious and usual candidates though not necessarily good ones, are as follows:

$$t_1 = \frac{1}{m} \sum_i z_{i1},$$
$$t_2 = \frac{1}{m} \sum_i z_{i2},$$

and

$$t_2 - t_1 = \frac{1}{m} \sum_i (z_{i2} - z_{i1}).$$

This last statistic is appropriate to additive effects. Our task is then to examine the variability of this statistic *over the randomization possibilities.* We wish to understand this variability, the expectation, variance and indeed, the whole randomization distribution of the statistic.

In writing the expression for $(t_2 - t_1)$, I have implied, of course, that a number y_{ijk} can be written as the sum of 2 numbers, one depending on the sub-unit index (ij) and one depending on the stimulus index k.

To do this, is to bring in, albeit very innocuously, the assumption that there is no interaction of sub-units within units with stimuli, *an additivity assumption.*

May such an assumption be used without concern? I believe not. It is obvious, ab initio, that the ordinary naive analysis, whether it be by analysis of variance or by $C(\alpha)$ test with its infinite population plus random sampling assumption, is based strongly on the use of this assumption of absence of interaction of sub-units with stimuli. One will, of course, try to make this assumption more reasonable by choice of function of actual measurement that one looks at, what for decades was known as transformation, but now some wish to call "reexpression."

The function of the measurements for this crossover design used in various studies is very interesting and perspicacious:

$$\left(\frac{\text{Total Seeded 2}}{\text{Total Unseeded 1}} \right) \times \left(\frac{\text{Total Seeded 1}}{\text{Total Unseeded 2}} \right)$$

where 1, 2, refer to sub-units. This in terms of the design random variables is equal to

$$T = \frac{(\Sigma \delta_i y_{i22})}{(\Sigma \delta_i y_{i11})} \times \frac{(\Sigma (1 - \delta_i) y_{i12})}{(\Sigma (1 - \delta_i) y_{i21})}.$$

This criterion is very interesting, because it reflects a quite different idea of "additivity" of treatments and sub-units. It would be interesting to attempt to obtain some continuous distribution approximation to the randomization distribution of T. However, a "solution" to this is not needed in any case, because one can examine the randomization distribution of T, and then judge significance of the observed value by reference to this distribution.

The expression for T, in the above, in terms of $\{\delta_i\}$, indicates how one gets the randomization distribution merely by using the possible vectors $(\delta_1, \delta_2, \ldots, \delta_n)$. In fact, one can construct any function one wishes of the data and δ. I suggest that original ideas on different criteria one might use could be useful. It is certain that some attempt must be made to examine past data for nonadditivities *in a randomization framework*.

The background of conceptualizing the crossover experiment shows clearly that it is assumed that the imposition of a stimulus on one sub-unit does not have any effect on what one would observe with another sub-unit with the same or different stimulus. This seems to be essential, because otherwise one will have to introduce into one's modelling, for instance, two numbers y_{i11} and y^*_{i11}, in which the former is what happens with sub-unit $(i1)$, say, if stimulus 1 is on sub-unit $(i2)$ and the latter is what happens if stimulus 2 is on sub-unit $(i2)$. The existence of such effects will "mess up" the whole problem and any modelling of stimulus effects.

6 COMMENTS ON THE FSU STATISTICAL PROJECT

Here I shall give some reactions to the reports of statistical work done at the Florida State University of Phase I of the Santa Barbara Project II.

1. The idea of using "response surface methods" to get a better number for "yield" is interesting. I wonder, however, as does A. Court, whether this effort which does not take account of topography may be expected, on scientific grounds, to be fruitful. I was interested by Court's remarks that the assessment of rain yield *over an area* is a widely studied problem of hydrography. One can envisage a wide variety of methods, e.g., of constructing a polygonal map from the rain gauge measurements and then using the area under it. Some such method has more appeal to me than "response surface methodology."

It is interesting and relevant, perhaps, to future experiments that this new method of assessing yield had nearly unit correlation with the simple arithmetic average.

2. The effort on multivariate approaches to summarizing the huge observational record of the experiment is unlikely, I surmise, to lead to any deeper understanding. The question at issue is whether seeding gave a change, e.g., increase, in yield over the target area. I use an analogy with an ordinary field plot experiment; examination of the variability *within* a plot is not likely to lead to improved understanding of differences, potentially related to treatments, *between* plots. But in giving this view, I must emphasize that it represents only a surmise.

3. The search for covariates is surely rational. I would hope, however, that physical theory would provide a reasonably short list of candidate covariates.

4. I am not, at least at first thought, inclined to favor the weighted analyses that were done. Unequal weighting alters, in a sense, the variable that one is trying to understand.

5. One will certainly use a linear model with covariates, and associated statistical methodology as a first routine attack, as was done. I am of the opinion that all the ideas tried by the FSU group had to be tried. They do, of course, depend on assumption of random independent error and this may well be questionable, though the data are too diffuse for any analysis with reasonable sensitivity, I opine. The upshot was that there was found no evidence of a direct effect of cloud seeding on target area precipitation (FSU Statistics Report No. M467, p. 27).

6. This conclusion differs from that of North American Weather Consultants. I tend towards acceptance of the FSU judgment. However, an FSU report says it reaches "near agreement with the conclusion of Elliott and Thompson" of an increase due to seeding of near 50%.

7. The frequent occurrence of very small F ratios in analysis of variance directed towards detecting a seeding effect must surely cause discomfort. The first surmise when this occurs is the standard assumption of independence is not correct. It would be interesting to determine what happens with randomization test procedures. Perhaps, indeed, seeding had effects on the "control" area.

8. I was interested by and am in accord with the view that "Multivariate analysis does not appear to be a likely way to improved future experiment design and analysis," which was expressed in an FSU report. Over and above technical problems resulting from incompleteness of data, multivariate analysis methods tend to take a global look at all possible (linear combinations) of response or yield variables. A yield function that does exhibit treatment effects will be "swamped out" in a global test by all those other functions that do not show an effect.

7 COMMENTS ON OTHER PRESENTATIONS AT FSU

1. I regard the general discussion of philosophical ideas and the general weather modification area given by St. Amand as very useful. St. Amand says, in a nutshell, weather experiments must be based on what is known about the physics of the system. I was impressed by his examining the question of whether the stimulus (Ag I, say) actually reached the "flying pig," to use my analogy. I accept St. Amand's written and oral presentation that seeding done under very particular circumstances has remarkable and undoubtable effects. The fact that the necessarily naive "experiments" have not shown effects of seeding serves mainly to show that confirmatory experiments are very difficult to design and perform.

2. I incline to the view that Neyman's keenness of $C(\alpha)$ tests is misplaced, for reasons I have given earlier. I do applaud, however, the efforts of Neyman and his collaborators on data analysis, with the sole proviso that we simply do not know how to "water down" reasonably the significance of effects that we have found after very extensive data search. I also applaud the Berkeley symposium (1965) effort.

3. I do not attach high value, vis-a-vis weather modification, to work on the mathematics of testing hypotheses on various mathematical distributions, though again this may be useful with regard to suggesting test statistics that one will examine via the experiment randomization distribution.

4. I found myself impressed with Elliott's overall presentation and his ideas on the new Colorado experiment that is to be done. My only caveat is that the planning of the study must contain a protocol, rather exact, of the assessment procedures that will be used when the experiment is regarded as *confirmatory*.

5. I am in agreement with the views expressed by Gabriel (1979).

6. I am largely in agreement with the material of the oral presentation of Court. I note, indeed, that many of the views I express were given by Court himself (Court, 1965).

8 CONCLUSION

Modification of weather could surely be a process of tremendous value to mankind. The work of the past three decades or so has shown that the area is of huge difficulty. It is not enough to randomize. This is just one small, but necessary, component of the whole experimental attack.

I find that I have considerable malaise with the numerous conference reports over the years, at intervals of two years or so. They seem to me to have added little. I incline to the view that weather scientists still do not understand the nature of the randomized experiment, the role of experimental units on which, of course, this randomization is based, the necessity of defining usefully the "flying pig" to which a stimulus is being given and the necessity of defining the "yield." I do not see these problems being addressed at all adequately. The whole training of physicists, I suggest, militates against the appreciation of these necessary ideas. How otherwise can one explain the continuing utter naivete of using ground burners that spew Ag I up into the sky somewhere (!)?

I do not wish to imply *at all* that the whole area of research should be terminated. I do suggest, however, that there is a desperate need for a close collaboration of cloud physicists and statisticians, in which each group makes very arduous attempts to understand the other, and in which each group attacks the ideas of the other without fear or favor. I think interaction of physicists and statisticians has been unsuccessful. Much of this failure, but not all, I attribute to the complexity of the problem, which is far deeper in many ways than any standard problem of physics, in the study of which there is really no experimental unit variability problem. Future experiments must be much more carefully planned than those of the past. There is need for huge input of measuring devices. I incline to the view that experiments should be confined to well-defined single clouds for the immediate future, so that the physics of the underlying processes can be worked out.

On the matter of data analysis of past studies, I believe that complexity of the problem has now been thoroughly uncovered, and mere "hunting" in past data of such uncertain status for significant relationships is most unlikely to produce repeatable causal relationships.

I am skeptical because I judge that the uncertainties of the nature of the experimental unit, of the nature of the delivery of the stimulus, and of the reliability of yield measurement are so great. I incline to the view that such activities do not advance our knowledge. The best statistical methods we know cannot overcome basic defects in the data. This is not, at all, a denigration of statistical methods, because everyone uses them. The "bottom line" is always data collection, analysis, and interpretation.

I have final remarks. The area is very difficult. I am impressed by the work of cloud physicists who are trying to get the science clear. I am not impressed by the comparative experiments I see. I think statisticians have oversold randomization because it cannot overcome basic defects of experiments.

BIBLIOGRAPHY

Court, A. (1967). Randomized cloud seeding in the United States. Proceedings of the Fifth Berkeley Symposium on Mathematical Statistics and Probability 5, 237-252.

Cox, D.F. and Kempthorne, O. (1963). Randomization tests for comparing survival curves. *Biometrics 19*, 307-317.

Fisher, R.A. (1935). *The Design of Experiments.* Oliver and Boyd, Edinburgh.

Gabriel, K.R. (1967). The Israeli artificial rainfall experiment. Statistical evaluation for the period 1961-65. Proceedings of the Fifth Berkeley Symposium on Mathematical Statistics Probability, 91-113.

Gabriel, K.R. (1979). Some statistical issues in weather experimentation. *Comm. Statist.-Theor. Method A8*, 975-1016.

Gabriel, K.R. and Federa, P. (1969). On the distribution of statistics suitable for evaluating rainfall simulation experiments. *Technomtrics 11*, 149-160.

Hanson, M.A., Bach, C.L. and Cooley, E.A. (1976). Bibliography of Statistical and Meteorological Methodology in Weather Modification. Florida State University Statistics Report No. M388.

Kempthorne, O. (1952). *The Design and Analysis of Experiments.* Wiley, New York, reprinted 1973, Krieger, Huntington, N.Y.

Kempthorne, O. (1955). The randomization theory of experimental inference. *J. Amer. Statist. Assoc. 50*, 946-967.

Simpson, J. (1974). Weather modification: Where are we now and where should we be going? Convective cloud modification. 4th Conference on Weather Modification of the American Meteorological Society.

Statistical Task Force. (1978). The Management of Weather Resources. Vol. 2. The role of statistics in weather resources managment. Tukey, J.W., Brillinger, D.R. and Jones, L.F. Department of Commerce, Washington, D.C.

Sutter, G.J., Zyskind, G. and Kempthorne, O. (1963). Some aspects of constrained randomization ARL 63-18. Office of Aerospace Research, Wright-Patterson Air Force Base, Ohio.

Wilk, M.B. (1955). The randomization analysis of a generalized randomized block design. *Biometrika 42*, 70-79.

Limitations of Statistics in Weather Modification

Arnold Court

Department of Geography
California State University
Northridge, California

Abstract

Analyzing a weather modification experiment requires statistical skill in testing physical hypotheses formulated by meteorologists, but not in proposing any new ones to investigate. Randomization bias often is present, and must be studied. Methods of estimating areal precipitation, and of allowing for errors in rainfall correlation should be considered.

Statistical analysis is essential for the success of any experiment in weather modification, but such analysis has very strict limits. Essentially, it should be restricted to evaluating whatever accomplishments are claimed for the experiment, or any hypotheses adduced by meterologists to explain the results. All such claims, hypotheses, and conjectures must be formulated jointly by meteorologists and statisticians, so that on the one hand they possess physical validity while having practical and operational value, and on the other hand they are amendable to statistical test using the observations to be available.

All hypotheses concerning the outcome of any experiment, especially one in weather modification, should be formulated in advance, together with procedures for collecting and analyzing the data to test them. Unfortunately, in past experiments few such hypotheses have been presented. Hence the analyses, and especially the reanalyses of the type discussed in this workshop, must develop the hypotheses de novo. This requires great care, to insure that such a posteriori formulation of hypotheses does not ignore the basic intent of the experiment, and focuses instead on some other question. If that new question has scientific validity and practical importance, it may be worth investigating, but first the primary purpose should be established and tested. Cloud seeding is a complicated business, and hypotheses concerning it should be formulated by, or in consultation with, meteorologists experienced in it.

In any weather modification evaluation, the major purpose should be to *test the claim* of the experimenter that he will have accomplished something. Usually that accomplishment is more water in some form, but it can be less hail, or less wind in the case of some hurricane experiments, or more sunshine in those activities intended to dissipate stratus clouds, or produce holes in them, which is apparently a rather successful operation. The usual goal of more water—rain or snow—in most experiments is over a defined area, such as the drainage area of a river or lake or reservoir, but may be an extensive agricultural area where farmers rely on rainfall rather than irrigation.

Sometimes, however, the experiment is more complex and seeks to develop procedures for "targeting" the precipitation increase assumed to result from treatment. Then the "more water" is not over a stationary target, but over a variable area defined by the winds at various levels from the ground up to cloudtop. Objective ways to define such target areas must be found, without regard to actual precipitation. Any rainstorm must have a region of heaviest precipitation, which may or may not have been caused by the treatment under evaluation.

Except in such cases, the statistical evaluation of a weather modification experiment should not be concerned with the actual mechanics of the treatment. Whether the airplane flew at the right level, or whether the nucleant from generators on the ground was properly dispersed by the low-level winds, or whether nucleant was even produced, are not statistical problems. The basic question is whether the intended result was achieved.

Ronald Fisher's classic example of experimental design presented a methodology for determining if a certain woman could tell whether the milk or the tea was first placed in the cup. He made no effort to investigate *how* she reached her conclusion, merely to establish *whether* she could. Statisticians evaluating weather modification experiments should follow the same course.

Randomization is a *necessary but not sufficient* criterion for a successful experiment. Some statisticians seem to endow randomization with magical properties, and mutter dark suspicions about experiments if the treated and untreated periods do not have equal precipitation outside the target, away from the presumed effects of treatment. "Some problems with randomization" were suggested because more rain was indicated for the treated days, even without treatment, than on the untreated days of the Santa Barbara project. Professor Neyman has shown that in the Whitetop experiment in southern Missouri, 15 years ago, the treated days actually had less rain than the untreated ones, even before the treatment started. A similar result was found on the San Juan project in southwestern Colorado early in this decade: the treated days had naturally a little bit less potential than the untreated ones, but the difference was not significant.

Such differences are to be expected. Randomization of a finite number of days into two sets does not guarantee that each will have exactly the same amount of rain—especially when the two sets have far less than a hundred days each. The first step in any evaluation, therefore, is to use independent data of some sort to establish the differences in natural conditions on the two sets of days. For this, precipitation measurements from outside the target are useful, but so are atmospheric conditions not likely to be altered by the treatment: moisture content, temperature, stability just upwind from the operational area.

How much water fell over the intended target area during the treated periods, and how much during the control periods, must be estimated in most weather modification experiments? Estimation of areal precipitation is the central problem in the field of hydroclimatology, in which I teach a course each year. And it is far more complicated than most statisticians assume. Before embarking on a weather modification evaluation, the statistician should consult with a competent hydrologist, and review the many techniques that could be used. Properly, of course, the method of estimating areal precipitation should have been established a priori, as part of the experimental protocol. But rarely is this done.

A storm will leave a mound of water behind as it moves across the landscape. Over a small level portion of its total extent, this mound conceivably can be approximated by a plane, perhaps horizontal, more likely inclined. An individual shower often produces a cone of rainfall, which is swept into an elliptical mound as the storm waxes, moves and wanes.

Over mountainous terrain, such as in Santa Barbara county, no such simple model will suffice. Certainly the unweighted average of all gages provides a poor estimate of the average depth of water over the target. The available methods involve weighting each gage reading by the area represented by the gage, or the elevation interval in which it lies, or by other criteria. The development of such procedures during half a century should be considered before one blithely adopts some simple-minded averaging procedures.

Likewise, the vast literature on rainfall correlations should be studied before such correlations are used in some hocus-pocus factor analysis. Ronald Fisher was the first, so far as I know, to study the decrease with distance in the correlations of the rainfall amounts at two gages. His 1922 paper discussed the correlations of annual rainfall at Oxford with that at stations throughout western Europe, as far east as Odessa. In the ensuing half-century correlations of annual, monthly, and daily rainfalls have been studied. In general they decrease with distance, reaching zero (for annual values) at 1,000 km or so, becoming negative, then becoming positive at another 1,000 km. Such behavior is more pronounced east-west, in the direction of storm motion, than it is north-south.

Many students err gravely in assuming that the correlation is unity at zero separation: actually, it is somewhat less, perhaps .90 or .95. Two adjacent rain gages will differ somewhat because of slight differences in exposure, construction, operation, and interpretation. Part of the disagreement between gages 10 or 25 or 500 km apart arises from measurement errors, and not from distance.

If the form of the correlation decay function were known—if certainly is not a simple negative exponential, nor a half-normal, but may be a kind of damped wave—it could be fitted to the data and extrapolated back to zero separation to obtain an estimate of the instrumental error. (A half-normal was used 15 years ago to estimate radiosonde errors.) Until this can be done, we could assume that the true correlation at any distance is 5 to 10 percent more than the observed value.

All these difficulties with estimation of areal rainfall, and inter-station correlations, render difficult any interpretation of principal components analysis—which is a manipulation of a matrix of raw correlations. Certainly such analysis has very little to do with the basic problem: did more rain fall during and after treatment than would have otherwise?

Statistical evaluation of a weather modification experiment is primarily a test of the operator's success in accomplishing his objective, which must be ascertained at the outset. If it is to produce more precipitation over a target, the various methods of estimating such areal precipitation must be considered before one of them is adopted. So must the many studies of the statistical behavior of precipitation, and of the correlations between gages various separations. *Whether* the treatment resulted in detectable differences, but not *how* it worked, is the statistical question.

Comments on the Reanalysis of the Santa Barbara II Cloud Seeding Experiments

K. Ruben Gabriel

Department of Statistics
University of Rochester
Rochester, New York

Abstract

Precipitation data from convective bands is subject to serial dependence. The only valid method of testing them for significance is that of rerandomization (permutation tests). ANOVA tests *might* perhaps serve as approximations. All significance tests of a variety of subhypotheses in a reanalysis, such as this, are subject to the problem of multiplicity and any "significance" found must be suitably diluted. The use of covariates seems essential, but the only good covariates are precipitation data in nearby areas and these are suspect of being dynamically affected. If so, are there no valid covariates? And does that mean that we cannot continue to experiment with precipitation? Various sophisticated data summarizations have been shown to make no improvement over simple average rainfall amounts. That includes principal components, even though the second and third PCs do reveal systematic topograhic patterns. An interesting finding of several significant t's observed simultaneously with a nonsignificant T^2 is explained by detailed consideration of rainfall variability and by the finding that seeding apparently affects all subareas equally. This illustrates the usefulness of multivariate techniques in exploring data rather than in testing for significance. Some comments are made on the interaction between statisticians and cloud seeders and physicists.

1 INTRODUCTION

The main part of my remarks is concerned with the statistical reanalysis of the Santa Barbara II experiment by statisticians at the Florida State University under the leadership of Dr. Ralph Bradley.

This is an impressive effort which adds to our understanding of the methodology of rainmaking experiments. Though many of its findings are negative, this is also useful as it indicates avenues which one need not explore again in the future. The thoroughness of the Florida group's work is commendable and will set a standard for future analyses.

I will comment on a number of aspects of the Florida reanalyses, and refer the reader for more detail to a forthcoming paper of mine (Gabriel, 1979a).

2 TESTS OF SIGNIFICANCE

Units. The Santa Barbara II project was randomized in convective bands. These were supposed to be meteorologically meaningful and homogeneous units, unlike fixed time periods, but their use entailed considerable additional work in obtaining measurements and it introduced extra uncertainty and variability (Bradley et al., 1980). Also, no historical data on such units were available for reference or for choice of covariates. One would have to have very strong reasons for preferring such complex units rather than more simply defined and easily measured ones such as days.

Use of convective band units entailed a further difficulty. Precipitation on successive bands was recognized as likely to be serially dependent. The Florida statisticians were aware of this possibility, but they do not seem to have analyzed its extent—such a study would be a welcome addition to the reanalysis. What is more important, however, is that the occurrence of serial dependence violates one of the important assumptions of most parametric statistical analyses—independence of observations. In effect, it may invalidate all the Gaussian model tests of significance run by the Florida group: These tests can therefore be regarded only as descriptive statistics and the P-values ascribed to them by Gaussian theory must not be considered strictly valid.

Randomization Tests. The obvious way to deal with significance tests on such data is to use rerandomizations (permutation) tests as originally proposed by Fisher (1935). For such tests "the significance level has a *known* distribution under the hypothesis, this being guaranteed. . .by the randomization. . . not dependent on the properties of the units actually used" (Kempthorne, 1975, p. 322). Any statistic could be chosen for testing, its P-value would be determined by reference to rerandomizations which mimic the actual randomization used in the project. Cox and Kempthorne (1963) gave a very clear exposition of this method of testing and it has been used on several occasions with weather data (see for example Gabriel and Feder, 1969; Elliott and Brown, 1971), when a sample of the possible rerandomizations was used and the P-value defined as the frequency with which the actually observed test statistic was exceeded.

I have considerable doubts about the value of parametric significance tests for any weather modification experiment; we rarely know enough about the underlying stochastic behavior of precipitation. Parametric tests involve uncertain assumptions about distributions of natural and seeded rainfall and postulate independence of observational units. Their validity, be they normal ANOVA F-tests, $C(\alpha)$-tests or other variants, is as doubtful as the unverified assumptions on which they are based.

The following excerpts from Volume II of the Weather Modification Advisory Board's report are relevant:

> The device of judging the strength of evidence offered by an apparent result against the background of the distribution of such results obtained by replacing the actual randomization by randomizations that might have happened seems to us definitely more secure than its presumed competitors, that depend upon specific assumptions about distribution shapes or about independence of the weather at one time from that at another. The additional computing costs should now not be excessive. The precise null hypothesis for classical rerandomization analyses is, of course, that of exactly no effect rather than of a collection of effects about which we are indifferent, a difficulty arising for almost all methods of analysis. On balance, we recommend using such a rerandomization analysis—at least as one main alternative for the most crucial question—in any weather modification experiment.
> (WMAB Statistical Task Force, 1978, Section 15.)

> Essentially every analysis not based on *re*randomization assumes that the days (or storms) behave like random samples from some distribution. Were this true in practice, weather could not show either the phenomenon of persistance (autocorrelation) or the systematic changes from month to month within a season which we know do occur. In an era when rerandomization is easily accessible, and gives trustworthy assessments of significance, it is hard indeed to justify using any analysis making such assumptions.
> (WMAB Statistical Task Force, 1978, Section 25.)

I hope that the more important statistical tests carried out in the Florida reanalysis will be complemented by rerandomization evaluations of their significance.

Gaussian Approximations. One wonders how far the use of Gaussian theory tests might be justified as approximation to rerandomization tests along the lines of Pitman (1937) and Welch's (1937) work. The methodology developed at Iowa by Kempthorne and his students might come into play here (Kempthorne, 1952; Kempthorne, 1975). The applicability of these approximations to weather data ought to be examined: Are rainfall variables sufficiently well behaved to justify the use of these approximations or do they exhibit such extreme variability that the approximations become poor?

Answers to these questions would tell us to what extent randomization arguments might validate the ANOVA type analyses carried out at Florida.

Some Further Remarks on Significance Testing. The Florida statisticians ran a large number of reanalyses of the Santa Barbara II data. Here is a clear case of multiplicity: He who runs 100 analyses should not be surprised if about 5 turn out to be 5%-significant. It would be easier for statisticians to stimulate significance artificially by multiplying the number of tests than for meteorologists to stimulate precipitation artificially.

The Florida reanalysis also included instances where seemingly positive results were subsequently explained away by introducing or omitting concomitant variables. This is the counterpart of the above—an artificial inhibition of significance. Sufficient diligence in searching among concomitant measurements is almost certain to show some that are confounded with the treatment and may "explain away" its effects.

I do not mean to suggest that the Florida statisticians should not have done all these many reanalyses (or that other statisticians should not have gone back to reanalyze earlier experiments in many interesting ways). What I do mean is that the "significance" attached to their various analyses must not be interpreted literally in the two-decision accept-reject framework. It would have been absurd to take the most striking findings of the massive reanalysis and consider its P-value in the same way as one had regarded the significance levels calculated in the initial analysis of the Santa Barbara II project. Come to think of it, the fact that this is a reanalysis already militates against any strict interpretation of "significance" in testing. Why were these data submitted to reanalysis? Why were certain aspects of the project of interest? It is only too obvious that the null-hypothesis probability of a reanalysis being α-significant is well above α.

Again, I do not intend this as a criticism of reanalyses, but only as a warning about how to interpret "significant" tests in reanalyses. The value of the Florida statisticians' work is in their exploration of possible patterns of seeding effects. They could not have been expected to provide confirmation; that is not the role of a reanalysis.

Some protection against the effects of multiplicity (but not of choice of experiments for reanalysis) might be provided by adopting a simultaneous inference strategy for a number of tests. One might choose from among parametric methods used (Miller, 1966, 1977), randomization methods (Miller, 1966, pp. 189-192, Gabriel, 1979a, Section 4.3) or conservative Bonferroni bounds. Parenthetically, I would remark that statistical dependence of different tests does not invalidate most simultaneous inference strategies (though it makes Bonferroni bounds unduly conservative) since these strategies incorporate adjustments for dependence between the various tests included. (See Gabriel, 1979a, end of Section 4.3.) What is more crucial than the choice of method, however, is the delineation of the family of hypotheses to be tested, (including statistics to be used). Any simultaneous inference strategy is valid only within that family and not beyond it. In a free-wheeling exploration of data, the number of possible analyses is likely to be too large to delineate and count. Indeed, many of the analyses will be suggested by earlier analyses of the same data (the succession of Florida reports well illustrates this) and could not have been enumerated at the outset. Nor is it appropriate to enumerate only the hypotheses that were actually tested, since the real multiplicity includes also all hypotheses that *would* have been tested if the data had shown different patterns. Gauging this multiplicity is clearly an impossible task, just as it does not seem feasible to appraise the effect of selecting a project for reanalysis.

My remarks on multiplicity are essentially negative. I think we must be aware of the problem and beware of its obvious pitfalls, but unfortunately I do not think the available strategies of multiple comparisons address the main problem and provide an adequate solution. What we can, and must, do is to insist that any exploration of data should report the number of analyses that were considered and run.

3 COVARIATES

An important part of the Florida reanalyses was concerned with the search for covariates, pooling meteorological (Gleeson, 1977) and statistical (Bradley et al., 1978) expertise. A number of covariates were found to correlate highly with target precipitation. The most striking was control area precipitation which had almost as high correlation ($r^2 = .548$) as all the other meteorological covariates together ($R^2 = 0.597$). (Inclusion of the control area with the other meteorological variates raised the correlation to $R^2 = 0.712$.)

These seemed to be very encouraging results in the search for effective covariates, allowing a reduction of 55-71% of the experimental variance. However, it was later suspected by meteorologists that cloud seeding might have dynamically affected all these concomitant observations, both the *upwind* control area and the local atmospheric observations (Bradley et al., 1980). It was therefore recognized that all these observations might be invalid as covariates for analysis of seeding effects. What seemed excellent covariates on the basis of a priori meteorological considerations and statistical correlations, now appeared to be irrelevant, and possibly misleading, in the analysis of a cloud seeding experiment.

This is a most crucial issue for weather experimentation. The need for good covariates has been repeatedly emphasized, both at Florida and in many other places. If the conclusion from the Santa Barbara II reanalysis is that all the obvious concomitant measurements are invalid as covariates, including, most crucially, nearby control areas, then we shall be forced to look for other covariates—either concomitantly measured pretty far away from the target, or measured earlier, i.e., predictors. As best we know, any such covariates, though valid, will have much lower correlation with target precipitation and will do little to reduce its large natural variability. In effect, we would be back to single area experiments without controls or covariates.

If, indeed, that is the conclusion, then experiments would have to continue for 20 or more years to have a reasonable chance of detecting seeding effects (Schickedanz and Huff, 1971). Such experiments cannot be carried out in a democracy, and they go well beyond the 10-year efforts envisaged by the Weather Modification Advisory Board (1978, Volume I). If the meteorologists make such a strong case for dynamic effects of seeding that local or nearby concomitant observations are invalidated as covariates, then it is incumbent upon us, as statisticians, to report that confirmatory randomized weather experimentation is not feasible at this time.

On the other hand, I am not sure whether the evidence for dynamic effects is at present sufficiently conclusive to warrant the invalidation of concomitant measurements, especially of nearby control areas. It is not for me to judge the physics evidence, but I believe it is by no means unequivocal. As to the statistical evidence, it hardly exists. There is a fair amount of data on downwind effects, much of it quoted in Neyman's (1979) present paper and more available in a study by Brier, Grant, and Mielke (1974). There is very spotty evidence of crosswind or upwind effects: Some of it is quoted in Neyman's paper (1979) amongst a multiplicity of analyses of Grossversuch III and Whitetop. At Santa Barbara II the ratios of (control area precipitation on target seeded bands) to (control area precipitation when target bands were unseeded) were mostly less than one. This does not provide *statistical* support to the suggestion that seeding had positive dynamic effects in the control area. A careful analysis of the first Israeli experiment (Wurtele, 1971) did not show appreciable crosswind effects and concluded that even if such effects had existed their influence would most likely have been to bias the experiment conservatively, i.e., to underestimate the effects of seeding on the target.

Unless we obtain more conclusive evidence of appreciable dynamic upwind and crosswind effects of cloud seeding, we may feel that concomitant measurements could serve as covariates after all. If so, weather experiments could be conclusive after about 5 years and are therefore feasible. And, by the same reasoning, one would do well to increase the experiments' sensitivity by using crossover designs (Moran, 1959, Gabriel, 1967, Schickedanz, and Huff, 1971). If crosswind control areas can be used as covariates, they can also be used as alternate targets.

The decision is one for the meteorologists, cloud seeders and cloud physicists. We, as statisticians, must point out to them that if concomitants cannot be used as covariates, we do not at present have alternative covariates which would allow weather experimentation to be conclusive within a reasonable number of years. We can also add our opinion(s) about the statistical evidence of seeding effects on nearby areas. But it will be up to the meteorologists to make the decision of whether the available concomitants are indeed so suspect as to be invalid and whether, as a consequence, confirmatory randomized weather experimentation should be suspended.

4. SUMMARIZATION

Means, and Response Surfaces. An interesting aspect of the Florida reanalyses is the failure of sophisticated methods of data summarization. Response surfaces were no better than simple averages and single principal components were considered of interest mainly because they reflected overall averages.

The message is clear and agrees with the experience of other workers: The use of complicated calculations and weights is unlikely to salvage the data. It is better to use simple methods.

Principal Component: Analyses. A number of multivariate analyses of Santa Barbara II data were carried out at Florida (Scott, 1978). Principal component analysis was applied to precipitation data at different stations, separately for control area stations and for target area stations. (The units were also separated into seeded and unseeded. We have no additional comments to offer on that separation, so we do not mention it further here.)

It was found in both areas that the first P.C. accounted for over 70% of variability and its loadings were positive with all stations. The second and third principal components accounted for about 12% of variability and showed a clear geographic pattern. This was demonstrated by the Florida analyses (Scott, 1978, Figures 1-6) and could be shown very convincingly by comparing the *h*-plots (Corsten and Gabriel, 1976) of the second and third PCs with maps of the control and target areas (Figures 1 and 2).

For the control area (Figures 1a, 1b) there is a very close correspondence between the h-plot configuration of stations and their geographical location. There are a few discrepancies which should be worth investigating. Thus, the h-plot marker for E4144 is much closer to those of S235, E7946 and S201 than one would have expected from the map: Could it be that the mountain range North of these stations makes their precipitations more similar? [Scott 1978, p. 14 speaks of an "orographic component."]

For the target area the correspondence between h-plot configuration (Figure 2a) and map location (Figure 2b) was less close, but careful inspection shows that h-plot clusters do correspond to adjacent stations on the map and the up-down h-plot direction corresponds roughly to a south-west to north-east map direction. In fact, all but one station fit in pretty well with this correspondence — the exception being station NB, which clusters with Santa Barbara coastal stations though it is geographically located well to the north of the coast. Again, one wonders if there is a special explanation for this deviation from the general pattern.

The more traditional displays of PCs by the Florida statisticians picked up the general north-south and east-west characteristics of the second and third PCs. However, these were separate displays for each PC. The h-plot has the advantage of jointly displaying two PCs and thus it may provide more insight into the configuration of the variables.

5 MULTIVARIATE ANALYSES: HOTELLING'S T^2

Another calculation presented in the Florida reanalyses is a multivariate comparison of seeded with nonseeded bands, the 12 variables being averages over stations of 12 sub-areas arranged in the following pattern:

$$
\begin{array}{cccc}
9 & 10 & 11 & 12 \\
5 & 6 & 7 & 8 \\
1 & 2 & 3 & 4
\end{array}
$$

(Bradley et al., 1980). The Florida statisticians were surprised to find that the multivariate T^2 test was not at all significant despite the significance of most of the subareas' t tests. This is certainly an unexpected finding which deserves detailed consideration.

Write S for 12×12 matrix of within sum of squares and products and let its singular value decomposition be

$$ S = Q' \Lambda^2 Q = \sum_{a=1}^{12} \lambda_\alpha^2 q_\alpha q_\alpha' $$

where $\underline{q}$ is the α-th row of Q and Q is orthonormal). Also write $\underline{d}$ for the 12 element vector of differences between the means of the samples. Then the Student t-statistic for the variable combination with weights $\underline{\alpha} = (\alpha_1, \alpha_2, \ldots, \alpha_{12})$, is

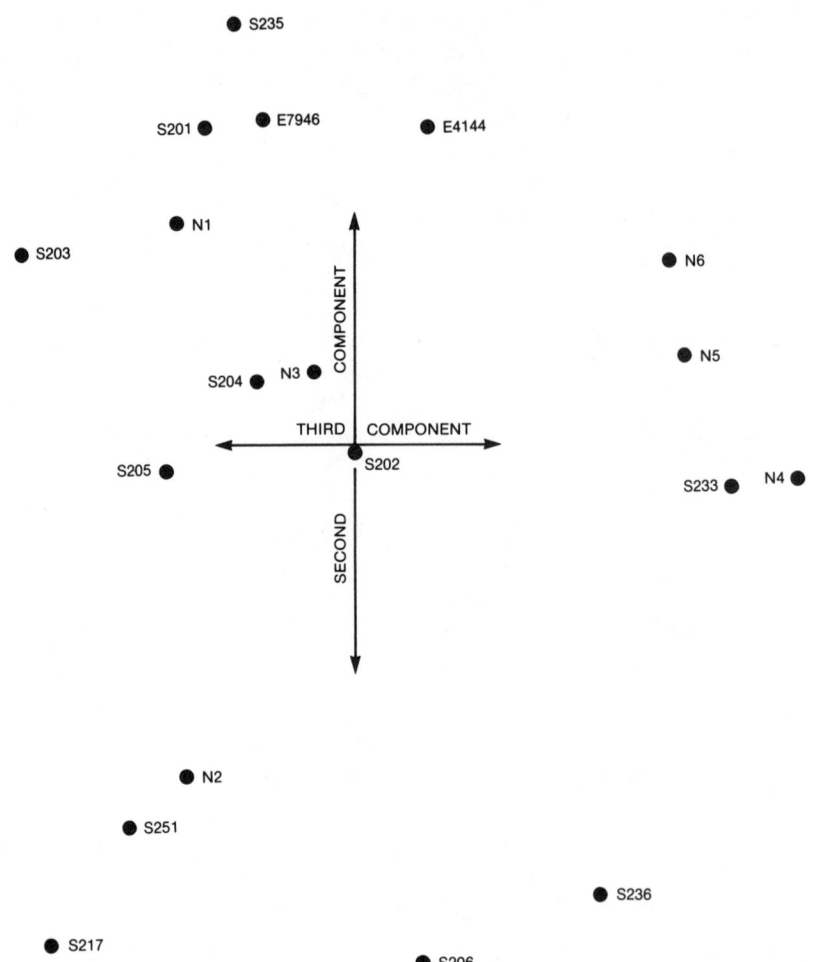

Figure 1a Central area station h-plot of 2nd and 3rd principal components.

$$t_{\underline{\alpha}} = C(\underline{\alpha}'\underline{d})/\sqrt{(\underline{\alpha}'S\underline{\alpha})}$$

for suitable constant C. Thus, for instance, for the ith variable, the t-statistic is

$$t_{\underline{u}_i} = C(\underline{u}_i'\underline{d})/\sqrt{(\underline{u}_i'S\underline{u}_i)}$$

where $\underline{u}_i$ is the vector with i^{th} element unity and all other elements zero. Defining

$$\underline{\beta} = \Lambda Q\underline{\alpha}$$

and

$$\underline{g} = \Lambda^{-1}Q\underline{d}$$

this becomes

$$t_{\underline{\alpha}} = C(\underline{\beta}'\underline{g})\sqrt{\underline{\beta}'\underline{\beta}}.$$

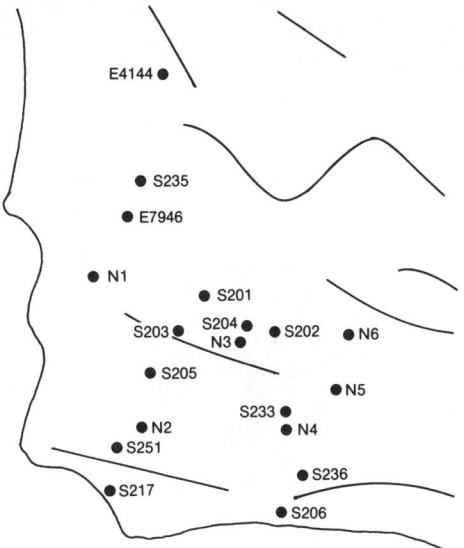

Figure 1b Central area stations locations.

As is well-known, for example, from S. N. Roy's Union-Intersection principle, Hotelling's T^2-statistic can be regarded as a maximum of the squared univariate t's. Thus

$$T^2 = \max_{\underline{\alpha}} t_{\underline{\alpha}}^2$$
$$= \max_{\underline{\beta}} C^2(\underline{\beta}'g)^2/g'g$$
$$= C^2 g'g$$

and therefore, for any $\underline{\alpha}$

$$t_{\underline{\alpha}}^2 = T^2 \cos^2(\Lambda Q \underline{\alpha}, g).$$

Under the null hypothesis of equal distributions, $t_{\underline{\alpha}}$ is a Student's t (central) for any $\underline{\alpha}$. However, T, the maximum of these statistics (in absolute value) does not have that distribution, but is stochastically larger than any individual $t_{\underline{\alpha}}$. The distribution T^2 is that of a maximum of squared t statistics. In the present case, the maximum T^2 was found not to exceed the individual $T_{\underline{u}_i}^2$'s by much, so it was not judged significant even though most of the individual $t_{\underline{u}_i}$'s were significant. In terms of the above development that means that

$$\cos^2(\Lambda Q \underline{u}_i, g)$$

cannot have been small for any $i = 1, 2, \ldots, 12$. In other words, g must have been close to proportional to each $\Lambda Q \underline{u}_i$ ($i = 1, 2, \ldots, 12$). Now,

$$\underline{u}_i' Q' \Lambda = (\lambda_1 q_{1,i}, \lambda_2 q_{2,i}, \ldots, \lambda_{12} q_{12,i})$$

but the twelve $(q_{1,i}, q_{2,i}, \ldots, q_{12,i})$ vectors are orthogonal to each other.

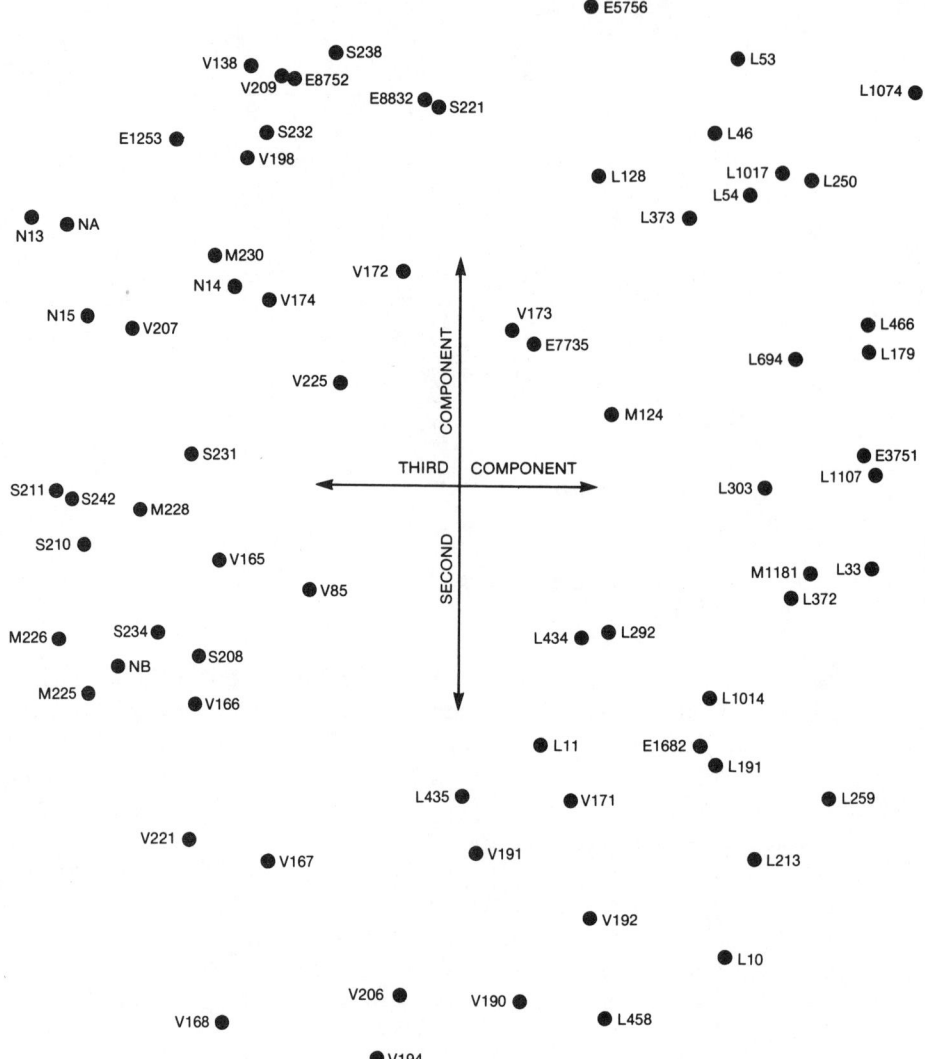

Figure 2a Target area station h-plot of 2nd and 3rd principal components.

Hence g can be close to proportional to all $\Lambda Q\underline{u}_i$'s only if $\lambda_2, \lambda_3, \ldots, \lambda_{12}$ are small compared to λ_1 and if

$$g = \gamma \underline{u}_1 + \underline{\epsilon}$$

for small $\underline{\epsilon}$. In the latter case

$$\underline{d}' = g'\Lambda Q$$

is approximately

$$\underline{d}' = \gamma \underline{u}'_1 \Lambda Q$$
$$= \gamma \lambda_1 \underline{q}'_1$$

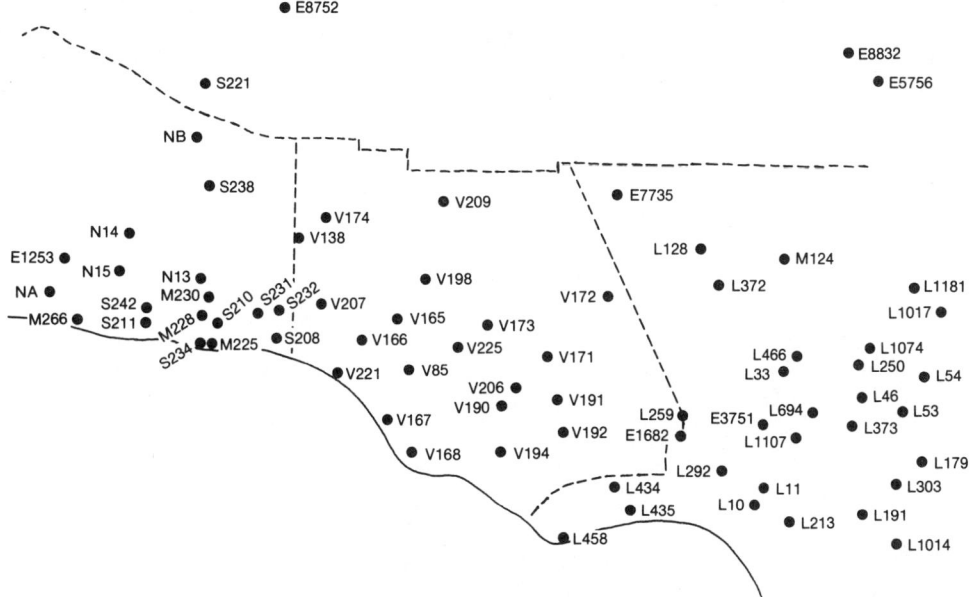

Figure 2b Target area station locations.

Thus, the situation could have arisen only if the within variation had a very predominant first principal component and the between sample difference vector is approximately proportional to that component. Considering what we have learned above about variability of precipitation, we know that there is a strong overall component of variation over the entire area (accounting for over 70% of variability). The observed phenomenon therefore suggests that the seeded/unseeded mean difference was almost entirely on that overall precipitation effect. One would infer that seeding may have affected overall precipitation over the entire target area but not the geographic distribution of precipitation (the next two components having been identified with the geographic location of the stations).

These inferences may be verified directly from the 12-variable data, as represented by the biplot planar approximation (goodness of fit 88.0%) or the three-dimensional bimodel (goodness of fit 92.7%). The biplot (Figure 3) shows a fairly tight sheaf of arrows for the 12 subareas, confirming their high intercorrelations. If further shows a clustering of convective band points to the left of the centroid (the plotting origin of the arrows). Towards the right the bands' points scatter more widely in what is very much like a 90° segment; this is typical of an asymmetric exponential type joint distribution with a concentration about the origin (represented on the biplot by the vertex of this 90° segment, i.e., immediately to the left of the points cluster).

Points for seeded convective bands are marked distinctly from points for unseeded bands. The scatter is too erratic to permit easy comparison of the two types of points, though it would seem that the seeded points are more frequent on the right half of the biplot, corresponding to higher precipitation values. Centroids and concentration ellipses have therefore been drawn onto the biplot

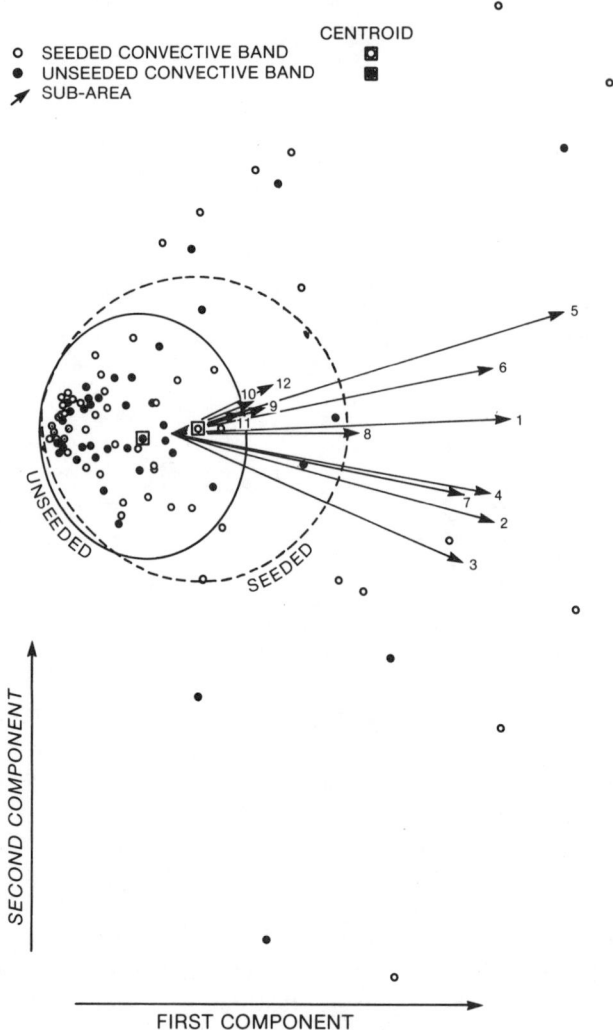

Figure 3 Biplot for 12 subareas 1st and 2nd components, with concentration ellipses for seeded and for unseeded bands.

separately for seeded and unseeded bands' points. The comparison is now unequivocal—along the horizontal axis (first principal component) the seeded bands have both a slightly higher mean (more average precipitation) and a considerably larger spread (more variable precipitation) than the unseeded bands. Along the vertical axis (second principal component) there is next to no difference in means but again seeded variability is a bit larger. For the third component (Figure 4) there seems little difference in either mean or variability.

Graphical inspection of the biplot/bimodel thus confirms our supposition that the seeded/unseeded difference is largely on the first principal component. It further shows that variability is also larger under seeding, a result which concurs with similar findings in Israel (Corsten and Gabriel, 1976).

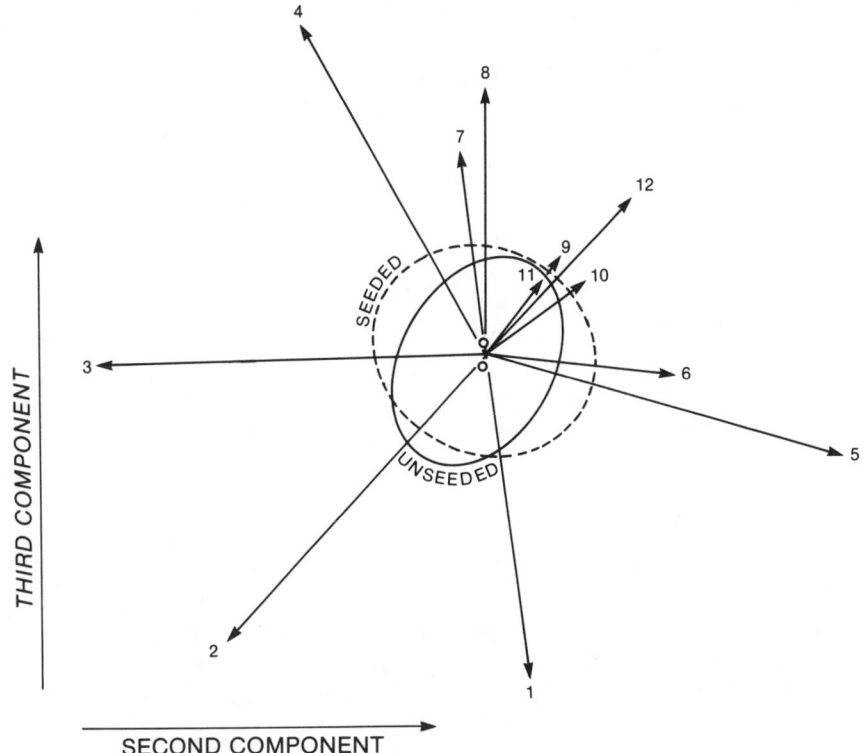

Figure 4 Biplot for 12 subareas, 2nd and 3rd components, with concentration ellipses for seeded and for unseeded bands.

It is interesting to note, by the way, that the biplot for the second and third components (Figure 4) roughly confirms the geographic pattern noted in the earlier analysis. For the Southern subareas 1,2,3,4, as well as for the middle subareas 5,6,7,8 the west-east gradient is roughly a 5 o'clock to 11 o'clock gradient on the biplot. Orthogonal to it is a south-north gradient (1,2,3,4 vs. 5,6,7,8). We also note that subareas 9,10,11,12 do not appear to conform to this pattern. Actually, this biplot also suggests that the seeded bands have rather more east-west variability than the unseeded bands, and less north-south variability. This again would seem to contradict the proposed dynamic effects which were supposed to have caused seeding effects to extend upwind, i.e., westward, and thereby presumably reduce east-west variability.

My reanalysis of these multivariate data shows that they provide a great deal of very interesting information about apparent seeding effects. Some idea of the support of these findings is provided by comparing the PC scores of the seeded and unseeded band by means of the Wilcoxon-Mann-Whitney test: $P = .03, .48, .40$ for the 1st, 2nd and 3rd and PCs, respectively. This test is valid by a rerandomization argument since bands were randomly assigned to be seeded or unseeded. No assumptions about the distribution of PC scores or of independence of bands is involved. However, it may well be argued that the statistic for testing was chosen out of a multiplicity of possible choices so that its significance is diluted. Indeed, we would not stress the confirmatory value of this analysis unduly.

After this detailed exploration of these multivariate data, I cannot agree with the Florida statisticians' pessimism about the about the use of multivariate analysis (Bradley et al., 1980). What is evident is that direct application of multivariate significance tests was unproductive. But multivariate exploration has been highly suggestive. Perhaps this is a lesson we should learn more generally. Multivariate methods may not be particularly effective for confirmatory calculations. Indeed, by the time research gets to the confirmatory stage one should be able to formulate one's null hypothesis sufficiently precisely to decide exactly what single variable it is to be tested on. However, multivariate analyses may come into their own in allowing data to be explored in much greater detail than is possible by univariate methods. This is not how we were taught (and have taught) multivariate analysis, but it may be no less true for that.

6 ON THE INTERACTION OF STATISTICIANS AND METEOROLOGISTS

I would like to conclude with a few general comments on the interaction between statisticians and meteorologists in the context of cloud seeding. This has gone on for more than 20 years and has generated an undue amount of friction without really providing the cloud physicist with clear advice on how to plan his experiments. (See Braham, 1979, for an excellent discussion. Also, Gabriel, 1979b.)

To begin with, let me voice a difficulty we statisticians have. Meteorologists and cloud seeders often state their "case" rather more positively than we are used to. They tend to sound as though they were familiar with the precise mechanism of rain formation and knew exactly how different kinds of seeding affect it. Sometimes, when one listens to their explanations and watches their displays of data, slides and film, one is tempted to be carried away with their enthusiasm. Be we statisticians are a dull lot, professional skeptics trained to hold onto the null hypothesis until the evidence overwhelms us significantly, so we prefer to treat many of these statements as working hypotheses, rather than as established facts. As such, we can respect these inspired guesses since they help to guide the cloud seeder towards further research and development.

If some of the "facts" cited by cloud seeders seem to need further investigation, this does not justify the alternative strategy adopted by some statisticians who search for "facts" by voluminous exploratory data analyses without any support from the physics. Independent forays by statisticians into cloud physics are unlikely to yield much that is meaningful. I am reminded, in a reverse situation, of hearing a very eminent physicist report to a group of statisticians on the novel statistical test he had derived. One of the younger statisticians got up to point out that use of the Neyman-Pearson lemma would have yielded a better test. The eminent physicist then asked for a reference to that lemma .... I wonder if the cloud seeding theories that statisticians have advanced are treated with more deference by meteorologists.

Cloud seeding is meteorology, not statistics, and we statisticians are unlikely to do much good to the advancement of knowledge in this field if we do not succeed in making our expertise serve as support for the cloud seeders' work. All the statistical brilliance in the world will not design a good rainmaking experiment if cloud physicists do not provide the substantive content. It is unfortunate that the history of the intereaction of statistics and meteorology contains so many unpleasant episodes. The departure of some eminent cloud physicists from this area seems to have ensued on unhappy interaction with statisticians. That is not a role our profession can be happy with.

Our role as statisticians cannot go beyond that of supporting the best expertise in the field. We cannot supplant it. When we statisticians succeed in driving out the most capable and enthusiastic research workers, we gain a Pyrrhic victory. The next experiment may be perfect in precision and design, but its substantive content will have been lost. We could have run it more cheaply by tossing dice.

Acknowledgment

I appreciate the readiness of Dr. Ralph Bradley and Dr. Elton Scott in providing me with the multivariate data whose analysis I have presented in Sections 4 and 5. Vassilios Klonias's careful computational work is gratefully acknowledged.

BIBLIOGRAPHY

Bradley, R. A.; Srivastava, S. S.; and Lanzdorf, A. (1978a). An examination of the effects of cloud seeding in Phase I of the Santa Barbara Convective Band Seeding Test Program. Technical Report M467, Department of Statistics, Florida State University, Tallahassee, Florida.

Bradley, R. A.; Srivastava, S. S.; and Lanzdorf, A. (1980). Some approaches to statistical analysis of a weather modification experiment. This volume.

Braham, R. R. (1979). Field experimentation in weather modification. *J. Amer. Statist. Assoc. 74*, 57-68.

Brier, G. W.; Grant, L. O.; and Mielke, P. W. (1974). An evaluation of extended area effects from attempts to modify local clouds and cloud systems. *Proceedings of the WMO/IAMAP Scientific Conference on Weather Modification.* Tashkent, USSR, 439-447.

Corsten, L. C. A. and Gabriel, K. R. (1976). Graphical exploration in comparing variance matrices. *Biometrics, 32*, 851-863.

Cox, C. P. and Kempthorne, O. (19963). Randomization tests for comparing growth curves. *Biometrics, 19*, 307-317.

Elliott, R. D. and Brown, K. J. (1971). The Santa Barbara II project. Downwind effects. *Papers presented at the International Conference on Weather Modification.*

Canberra, Australia, September 6-11, 1971. American Meteorological Society.

Gabriel, K. R. (1967). The Israeli rainfall stimulation experiment, statistical evaluation for the period 1961-65. *Proceedings of the Fifth Berkeley Symposium on Mathematical Statistics and Probability, Volume V, Weather Modification* (Lucien LeCam and Jerzy Neyman, eds.) Berkeley, University of California Press, 91-113.

Gabriel, K. Ruben (1979a). Some statistical issues in weather experimentation. *Comm. Statist.-Theor. Meth. A8*, 975-1016.

Gabriel, K. Ruben (1979b). Comment on Braham's paper. *J. Amer. Statist. Assoc. 74*, 81-84.

Gabriel, K.R. and Feder, P. (1969). On the distribution of statistics suitable for evaluating rainfall stimulation experiments. *Technometrics 11*, 149-160.

Gleason, Thomas A. (1977). Data summarization in a weather modification experiment: II Concomitant variables. Florida State University, Department of Statistics Report M419 (mimeographed).

Kempthorne, Oscar (1952). *The Design and Analysis of Experiments.* New York, Wiley.

Kempthorne, Oscar (1975). Inference from experiments and randomization. *A Survey of Statistical Design and Linear Models.* (J.N. Srivastava, ed.) Amsterdam, North-Holland, 303-332.

Moran, P. A. P. (1959). The power of a cross-over test for the artificial stimulation of rain. *Aust. J. Statist., 1*, 47-52.

Neyman, Jerzy (1979). Strong far-away effects of local cloud seeding. Manuscript.

Pitman, E. J. G. (1937). Significance tests which may be applied to samples from any populations. III. The analysis of variance test. *Biometrika, 29*, 322-335.

Schickedanz, Paul T. and Huff, Floyd A. (1971). The design and evaluation of rainfall modification experiments. *J. Appl. Meteor., 10*, 502-514.

Scott, Elton (1978). Data summarization in a weather modification experiment: III. Multivariate analysis. Florida State University, Department of Statistics Report M442 (mimeographed).

Weather Modification Advisory Board (1979). *The Managment of Weather Resources. Volume I: Proposals for a National Policy and Program.* Washington, D.C. US Government Printing Office.

WMAB Statistical Task Force to the Weather Modification Advisory Board (1978). *The Management of Weather Resources. Volume II: The Role of Statistics.* Report to the Secretary of Commerce. Washington, D.C. US Government Printing Office.

Welch, B. L. (1937). On the z-test in randomized blocks and latin squares. *Biometrika, 29*, 21-52.

Comments on the Discussion at the Workshop On the Statistical Design and Analysis of Weather Modification Experiments

Jerzy Neyman

Statistical Laboratory
University of California
Berkeley, California

1 INTRODUCTION

In a sense, the present "Comments" are one-sided. They reflect my own interests in weather modification developed during slightly more than a quarter of a century of studies, (Neyman, 1954). Also, these interests appear to be shared by a group of colleagues at our Berkeley Statistical Laboratory. To a degree our interests are narrow: they focus on the large scale atmospheric phenomena that accompany, or follow, the seeding of clouds intended to affect the precipitation, frequently those intended to increase the rainfall or snowfall.

Obviously this domain of study is highly interdisciplinary. It involves certain aspects of atmospheric physics which we try to learn. Also, it involves some mathematical statistics which we try to cultivate. Our efforts are directed towards novel methodologies adjusted to peculiarities of cloud seeding experiments and, at the same time, convenient to use. However, no claims are made that we reached the top! A review of our findings is expected to appear in the special issue of *Communications in Statistics*, (Neyman, 1979a).

Having admitted the one-sidedness of our interests, it is hardly necessary to emphasize my readiness to recognize and to respect the interests of other research workers, different from my own.

The comments presented fall under the following headings:

(i) Specification of my narrow interests in weather modification studies.
(ii) Need of randomization of the experiments.
(iii) Some peculiarities of precipitation modification problems that require special evaluation methodologies.
(iv) Addendum

2 NARROWNESS OF MY INTERESTS IN CLOUD SEEDING EXPERIMENTS

My interests in cloud seeding experiments are limited to effects on precipitation reaching the ground in specified areas ("targets") or in categories of areas (perhaps far-away areas at specified distances, etc. with respect to targets), effects that could hardly be ascribed to vagaries of randomization. A great variety of other weather modification problems, such as the problems of hurricanes, of physico-chemical processes in particular clouds, etc., are outside of the sphere of my active interests. However, I am fully aware of the importance of these problems. In particular, I am appreciative of the study of Dr. St. Amand.

To illustrate my interests in areal effects of seeding, a reference to the Whitetop experiment is likely to be useful. This experiment differed from many others in that it had a fixed experimental area, the "research area," a circle centered at the airport at West Plains, Missouri. Over the years, our interests focused on the apparent effects of seeding on rainfall within the Whitetop research area. On the other hand, the seed/no seed differences in precipitation in the day to day variable areas labeled "Chicago Plumes" or "Missouri Plumes" or "Out of Plumes" appeared less interesting.

At this point my attitude appears to differ from that of Professor Arnold Court. This is illustrated by Court's first sentence: "Statistical analysis is essential for the success of any experiment in weather modification, but such analysis has very strict limits. Essentially, it should be *restricted* [emphasis added] to evaluating whatever accomplishments are claimed for the experiment, or any hypotheses adduced by meteorologists to explain the results." Somehow, the emphasized word "restricted" bothers me. However, I feel somewhat comforted by Court's statement in his Abstract that "Randomization bias often is present [in cloud seeding experiments] and must be studied." My point is that our ultimate conclusion on the Whitetop experiment is our own hypothesis that "there must have been some difficulty with randomization."

This hypothesis is emphatically disapproved by Professor Braham who conducted the Whitetop experiment. The discussion of this point is to appear in (Neyman, 1979b).

Another argument against Court's "restricted" is brought out in my comment in the forthcoming issue of *JASA*. This is that Professor Braham's interests in the effect of the Whitetop seeding were somewhat limited. First they covered six hours beginning late morning. The latest I saw covered the subsequent five hours, up to 9 p.m. of the experimental day, for which period some significant apparent effects of seeding were found. While Braham appears to have stopped at 9 p.m., I find it too restricted and am interested in the seed/no-seed differences in rainfall on the 24 hour period, which was Braham's unit of randomization.

When Professor Court writes, "Professor Neyman has shown that in Whitetop Experiment . . . the treated days actually had less rain than the untreated days even before the treatment started . . .," he is right. This particular finding occured when we tried to verify a hypothesis offered by an atmospheric physicist to explain the unbelievably far-away effects of seeding. It is just these findings that caused our group to suspect that "there must have been some difficulty with randomization." For details, see (Neyman, 1977).

3 NECESSITY OF STRICT RANDOMIZATION

Here it is a pleasure to agree with Professor Court in his statement "Randomization *is a necessary but not sufficient* criterion for a successful experiment." In order to insure the reliability of an experiment, it is not sufficient to have it "randomized in the statistical sense" [this is a quote from an official report] but it also necessary that the randomization by "strict."

When emphasizing the "strictness" of randomization, I have in mind the following description published in (Neyman, 1977, p. 4717).

> With reference to precipitation augmentation, the essence of a randomized experiment is, briefly, as follows.
> First, "potential experimental period" (or "seeding opportunity") and the "response variable" are clearly defined. In the simplest case, the potential experimental period may be of fixed duration, say 24 hr from 0730 of a given day to 0730 of the next. In this case, the response variable might be the precipitation measured by specified gauges (that define the "target"), say from 0800 of the given day to 0800 of the next.
> A special organizational unit, to be called the "randomization center" (RC), must be established. At an appointed time before the beginning of a potential experimental period, the experimenter reports to the RC whether the approaching potential experimental period is suitable for inclusion in the experiment—that is, whether it is to become an "experimental unit." In the affirmative case, the experimenter communicates to the RC certain other information deemed important, such as the nature of the prevailing weather (type A, type B, or type C, etc.). In response, the RC provides the experimenter with a randomized decision, either a permissive "seed" or a categoric "do not seed." It is emphasized that the randomized decision must be communicated to the experimenter AFTER his declaration as to the approaching experimental unit, not before. In fact, it would be best to arrange that even the personnel of the RC have no advance information on the nature of the forthcoming randomized decision. Perhaps, a computerized random number generator could be adjusted for this purpose.
> All of these exchanges between the experimenter and the RC must be automatically recorded, perhaps by using teletype machines. The records at both ends must be consistent. The primary evaluation of the experiment must be based on *all the experimental units* (some seeded and others not) and no others, and it must use the originally defined response variable. The supplemental information about the type of weather ought to be used for stratification purposes and is useful by providing the experimenter with means to verify his ideas.

For quite some time, the reasons for the enumerated precautions involved in strict randomization were not mentioned in the cloud seeding literature. The novelty in this respect is represented by the following quotation from the voluminous report by (Elliot et al., 1976, p. 13, Chapter 6).

> Even the most sincere investigator may subconsciously misidentify a possible seeding opportunity if he knows whether the opportunity, if declared, will be actually seeded or merely kept as control.

The passage just quoted is from a report published in 1976, of which Arnold Court is coauthor. The many transgressions of "strict randomization" requirements could hardly be described as "subconscious." Here are two examples.

The first example refers to the widely publicized Israeli experiment I, described by Professor K. R. Gabriel and discussed in the Kempthorne essay. The original evaluation of this experiment (Gabriel, 1967) has the peculiarity that the whole period of its duration is composed of three subperiods, say I, II and III. The response variable in all the three subperiods is the 24 hr rainfall. However, during the subperiods I and III the 24 hr rainfall was measured from 8 a.m. of the given day to 8 a.m. of the next. On the other hand, during a longish period II, the 24 hr rainfall is measured in which Professor Gabriel calls "unconventional" manner, from 8 p.m. to 8 p.m. It is essential that no change in the timing of seeding is mentioned in the paper. The following paragraph and an illustrative figure are reproduced from (Neyman, 1971 pp. 15-17).

> While thinking about it I feel interested in the details of the second method of measuring day's rainfall. There are two possibilities and both seem unexpected. The situation is illustrated in Figure 3. [See Figure 1 of these Comments.] The quantity measured on the horizontal axis is time in hours. In the middle, there is symbolized a particular experimental day, from zero to twenty-four hours. Then there is indicated a hypothetical period of seeding and the "first" method of measuring this day's rainfall: from 8:00 a.m. of the given day to 8:00 a.m. of the next day. Various authors speak, and write, about the possible after-effects of seeding and, therefore, the inclusion of the night following a day with seeding does not seem objectionable. On the other hand, how to justify either of the two possibilities regarding the second method? If the rain measuring period starts in the evening of "preceding" day, then there seems to be a good chance of crediting the "given" day with precipitation affected by "yesterday's" seeding, this combined with the "loss" of rain of tonight likely to be affected by "today's" seeding. Again, if the rainfall of "today" is measured from 8:00 p.m. tonight to 8:00 p.m. tomorrow, similar questions arise.
>
> However, the most fascinating question is about the changes in the method, from method 1 to method 2, and then back to method 1. If both methods have some advantages then is it not interesting to see two evaluations performed uniformly, one using method 1 and the other method 2?

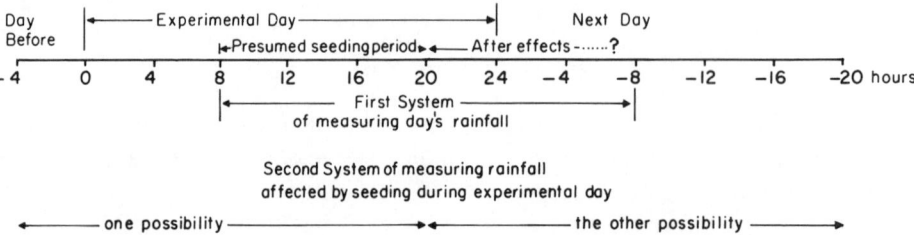

Figure 1 Guesses about measures of day's precipitation in Israeli experiment.

While the problem of the Israeli I experiment consists in the change in the definition of the response variable, the several evaluations of an important American experiment Climax I are open to question about selectivity of data used.

The cloud physicist involved in Climax I was Professor Lewis O. Grant. The experiment has been evaluated in cooperation with Professor P. W. Mielke and C. F. Chappell, the former student of Grant and Mielke. It is here that I have misgivings. The following is quoted from (Neyman, 1971, pp. 17-18).

> Figure 2 [Figure 2 of these Comments] summarizes the stratifications in the five consecutive papers concerned with Climax I, including the apparently unpublished M.S. thesis of Charles F. Chappell (references to 5 papers). The puzzling thing about these five papers is the variation of the total number of experimental days used. My regret is that, up to recently at least, the complete set of all the observations at Climax I, including dates, etc., has not been published.

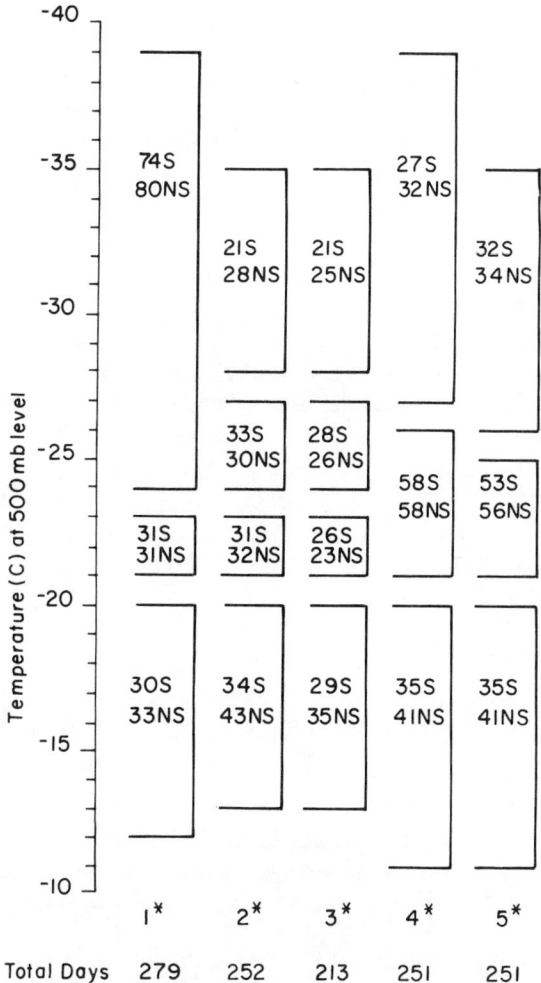

Figure 2 Variability in the number of experimental days used in evaluations of Climax I.

[*Professor K. R. Gabriel has taken strong exceptions to the interpretations of the Israeli experiment which were offered by Professor Neyman. He has written a paper entitled, "On Neyman's 'comments' regarding the first Israeli randomized rainfall stimulation experiment," which will appear in a forthcoming issue of* Communications in Statistics. *The Editors*]

4 PECULIARITY OF RAIN STIMULATION EXPERIMENTS

The peculiarity of precipitation modification experiments frequently overlooked in evaluations consists in the fact that the probability of the response variable being exactly zero is positive and that this probability can be affected by seeding. (Let us call this mechanism M_1.) Also, there is another mechanism, say M_2, through which the rainfall averaged "per wet day" can be altered by seeding. The two mechanisms M_1 and M_2 may be working "in the same direction," either tending to increase or to decrease the rainfall. However, they can also work in the opposite directions. For example, mechanism M_1 can increase the frequency of "wet days" while mechanism M_2 decreases the precipitation averaged per wet day. In these circumstances there is the possibility that the seed day rainfall averaged "per experimental day," whether wet or dry, can be equal to that on control days. In this case the overall effect of seeding can be exactly zero. However, many published evaluations of cloud seeding experiments ignore these points. The significance tests used are those of the so-called "null hypothesis" meaning the hypothesis that the seeding of clouds does not affect the *distribution* of rainfall.

It should be obvious that the question answered by these tests is not relevant to the problem of rain stimulation experiments. The relevant question is whether the seeding altered the mean rainfall. Through the functioning of the two mechanisms M_1 and M_2 this mean can be unaltered even though the distribution of rainfall on seed days is rather different from that on control days.

Now it may be useful to mention that the test frequently used in the evaluations of cloud seeding experiments presuppose that the distribution of rainfall per experimental unit is normal. Actually it is very far from being normal. As a result, the evaluations using analysis of variance tests can lead to unrealistic conclusions.

The above remarks are offered as comments on several presentations at the Tallahassee Workshop.

5 ADDENDUM

A substantial part of Professor Kempthorne's essay is given to the discussion of reasons for the noted slowness in the development of a reliable weather modification technology: partial failures of physical scientists and partial failures of statisticians. I agree with Professor Kempthorne on a number of points, but wish to mention a factor that appears to have escaped his attention. This factor is: misinformation stemming from official sources. One example must suffice.

The *Final Report of the Advisory Committee on Weather Control*, published in 1957, is in two volumes. The thin Volume I, containing a summary and recommendations, includes the following statement:

> No evidence was found in the evaluation of any project [commercial cloud seeding project] which was intended to increase precipitation that cloud seeding has produced a detectable negative effect on precipitation.

Not True. (See Neyman and Scott, 1961.)

Acknowledgments

Research for this report was supported, in part, by Office of Naval Research Contract ONR N 000 14 75 C 0159 and Department of the Army Grant DA AG 29 76 G 0167. All opinions expressed are those of the author.

REFERENCES

Elliott, R. D., Shaffer, R. W., Court, A., and Hannaford, J. F. (1976). *Colorado River Basin Pilot Project Comprehensive Evaluation Report*, Aerometric Research Inc., Goleta, California.

Gabriel, K. R. (1967). The Israeli artificial rainfall stimulation experiment: statistical evaluation for the period 1961-1965. *Proc. Fifth Berkeley Symposium on Mathematical Statistics and Probabilty*, Vol. V., eds. L. LeCam and J. Neyman, Univ. of Calif. Press, Berkeley, 91-113.

Neyman, J. (1954). Methodology of statistical evaluation of rain making operations. Mimeographed report to the Division of Water Resources, Department of Public Works, State of California, published in Bulletin 16, State Water Resources Board, Weather Modification Operations in California, Preliminary Report, August 1954, pp. C1-C24.

Neyman, J. (1971). Experimentation with weather control and statistical problems generated by it, in *Applications of Statistics*, ed. P. R. Krishnaiah, Amsterdam: North Holland Publishing Company, pp. 1-25.

Neyman, J. (1977). A Statistician's view of weather modification technology (a review). *Proc. Nat. Acad. Sci. 74*, pp. 4714-4721.

Neyman, J. (1979a), Developments in probability and mathematical statistics generated by studies in meteorology and weather modification, *Comm. Statist.-Theor. Meth. A8*, 1097-1110. See also first article, this volume.

Neyman, J. (1979b), Comment on Braham's Paper, *J. Amer. Statist. Assoc. 74*, 90-94.

Neyman, J. and Scott, E. L. (1961). Further comments on the final report of the advisory committee on weather control, *J. Amer. Statist. Assoc., 56*, 590-600.

Response to the Discussion

Ralph A. Bradley

Department of Statistics
The Florida State University
Tallahassee, Florida

Oscar Kempthorne, Ruben Gabriel and Arnold Court have provided written discussions of presentations at the Florida State University Weather Modification Workshop. In so doing, they have presented both their own views on the design and analysis of weather modification experiments and comments in regard to analyses presented by Srivastava, Lanzdorf and myself. Their interests and insights are appreciated. We are not in major disagreement with their views and respond in clarification on some issues.

Court has emphasized need for clear statement of experimental objectives, defined response areas, better experimental planning, and formulation of test procedures in advance of experimentation. We agree. We believe that he is concerned with the planning of a confirmatory experiment in the sense discussed in Volume II of the Weather Modification Advisory Board's report, although current state of the art may not justify such experiments. In retrospect, at least, Phase I of the Santa Barbara Convective Seeding Test Program must be regarded as an exploratory experiment. It was innovative in the use of convective bands as experimental units, a matter discussed in our paper. We have not found clear, specific statements of objectives. We have used the Phase I data to explore uses of various statistical approaches with the intent of investigating confirmation of the more promising ones with the Phase II data, although experimental changes in both techniques and experimental design will cause difficulties. We are in some disagreement with Court on other issues, his comments on randomization, his perceived role of the statistician (See Gabriel's concerns for the interaction of statisticians and meteorologists), his lack of concern for elimination of bad data arising for assignable cause, and his apparent objection to data exploration for possible new insights.

Kempthorne with his "pig flying in the sky," has outlined major difficulties in the problem area, difficulties of identification of suitable experimental units, measurement of response, large experimental variation, and appropriateness of statistical analyses. Gabriel has rather concentrated on the fourth of these and at some length. Both have noted the importance of good covariates unaffected by seeding. In our exploratory reanalysis of the Santa Barbara Phase I data, we concentrated on the development of a suitable response variable and an examination of possible means of control of experimental variation through consideration of available covariates, blocking, and model extensions for carry-over or positional effects. In so doing, we used a parametric approach as the most efficient means of data exploration, even though we expressed concerns for parametric assumptions. We reasoned that this approach should not be grossly misleading and interesting leads could be checked with the Phase II data.

One of our first concerns was for suitable summarization of raingage precipitation data attributable to an experimental unit. Use of a simple average seemed usual. As an alternative, we considered and have reported on, use of response surfaces to compute "precipitation volumes" as summaries of precipitations in both target and control areas for an experimental unit. It was found that precipitation volume was very highly correlated with precipitation mean and that little new could result from use of volume. A second approach was tried also. The "method of successive corrections," an algorithm said to be widely used in meteorology, was available through a computer program providing point precipitation responses and response contours. Contours so produced were very similar to those resulting from our reported response surfaces. Court has suggested yet another response index, a linear function of raingage readings, each raingage reading weighted by the surface area represented by the gage. Such an index may be suitable; we prefer the response surface method on intuitive grounds in that it provides a smoothing that is desirable and implicitly also gives more weight to observations representing larger surface areas. In any event, linear response indices giving reasonable weights to all raingage readings surely will be highly correlated with each other and with the simple mean of those readings.

Gabriel, Kempthorne and the cited Weather Modification Advisory Board report express concern about multiplicity of analyses of weather modification experiments. Kempthorne explicitly gives the standard formula showing that the probability of a Type I error is inflated as a number of *independent* tests of significance are performed. The problem exists; the issue is a difficult one, but one that we feel is overemphasized with highly *dependent* tests. Consider the response, precipitation mean, and one of our target areas, since results were consistent over our related target areas. The various analyses that we have presented could be regarded as a single analysis considering in a logical sequence various possible sources of variation affecting response (covariates, storms, positions in storms, carry-over effects, seeding); indeed, one could argue that they are exactly the sources of variation that should have been considered in planning the experiment and in a preplanned analysis of data. The situation is somewhat analogous to the analysis of a factorial experiment wherein certain higher order interactions may be tested, found nonsignificant, and their sums of squares considered to contribute to residual error. Questions arise in this analysis also but the procedure is widely accepted. There is a multiplicity of analyses problem. We disagree on its severity. Confirmation of significant results from exploratory analyses is needed.

Consideration of randomization tests seems necessary, if only to allay concerns. We have revised our paper to note that some randomization tests are planned.* Randomization tests based on independent samples of independent observations, without extreme values, may be well approximated by parametric tests, as recognized early by Pitman. However, dependencies of responses in the sequences of experimental units do cause concern. We shall be surprised if significance levels from the parametric tests are distorted badly, but a check is needed. Kempthorne has suggested randomization tests for covariance analyses; they seem possible only if the covariate measurements, which must remain associated with their respective experimental units, are not affected by seeding.

On a technical point raised by Gabriel on the multivariate analyses presented by Scott, the fact that T^2 may be regarded as the maximum of the squared univariate t^2's over possible linear functions of the variates seems less well known than the somewhat similar result for the squared multiple correlation coefficient. It may be demonstrated by direct maximization subject to appropriate constraint. In the Scott application of T^2, the value of T^2 is very close to the various values of t^2 for the variates individually.

We are now beginning consideration of the Phase II Santa Barbara data. In the Phase II experimentation, areal seeding replaced ground seeding for many of the convective bands. Because of concern for a persistence effect of seeding, all convective bands within a storm were either seeded or not seeded. We shall endeavor to confirm Phase I analyses with the Phase II data but the design changes made will cause some difficulties. We are disappointed that analyses to date have not led to major new insights to improved experimental design of future similar experimentation.

---

*Added in proof: These tests and additional comments on weather modification experiments are given in the following reference.
Bradley, R. A. and Scott, E. (1980), Perspectives from a weather modification experiment. *Commun. Statist.—Theory Methods*, **A9**, in press.

INDEX

Acetone burner, 21
Activation temperature, 17
Ammonium iodide, 21
Analysis of (co)variance, 41, 45, 98, 115, 141
Australian experiment, 24

Bowen, E., 24
Bradley, R. et al., 33, 67, 73, 103, 114
Braham, R., 126, 137
Bureau of Reclamation, 33

$C(\alpha)$ tests, 5, 92, 104
Catalysis theory, 17
Central limit theorem, 3
Characteristic functional, 2
Climax I, 135
Clouds
   air mass, 3
 cumuliform, 80
 cumulus, 21, 85
 frontal storm, 7
 orographic, 79
Cloud top temperature, 81
Coefficient of determination, 42
Colorado River Basin Pilot Project, 81, 110

Convective bands, 34
Correlation, 41ff, 62
Court, A., 105, 109, 132, 139
Covariates, 35, 79, 99, 117
Crossover design, 101

Design of experiments, 98
Distribution
    gamma, 3, 55
    J-shaped, 3
    lognormal, 55
    re-randomization, 55

Eignevalues, 66
Electricite de France, 2, 20
Elliott, R., 21, 23, 34, 49, 79, 106
Experimental unit, 5, 46, 48, 62, 92, 114

Florida Area Cumulus Experiment, 86
Flying pigs, 93, 139

Gabriel, K. R., 48, 96, 105, 113, 134, 136, 139
Glaciation, 83
Grossversuch III, 117

Hotellings T^2, 74, 113, 119, 141

Israeli I, 134

Kempthorne, O., 89, 114, 134, 139
Kolmogorov, A., 2

Langmuir, I., 14, 18
LeCam, L., 1
Little Egypt, 3

Moment generating functional, 2
Monte Carlo, 5, 38

National Hail Research Experiment, 6
Naval Weapons Center, 11, 39
Neyman, J., 1, 22, 92, 97, 110, 117, 131
North American Weather Consultants, 33, 61, 104
Nucleation, 18, 23, 80

Outlier, 1

Principal components, 40, 61, 113, 118
Project Hotshot, 24
Project Tignes, 20
Project Whitetop, 21, 85, 110, 117, 132

Radiosonde, 45, 81
Rain dancer, 13
Randomization, 5, 48, 91ff, 109, 114, 131
Regression, 41, 81
Response surface, 39, 118, 140

Santa Barbara Convective Seeding Test
   Program, 33ff, 55ff, 61ff, 87, 103, 113, 139
Sierra Cooperative Pilot Project, 87
Silver iodide, 17, 21, 90ff
Smith, E. J., 7
Saint-Amand, P., 11, 34, 95, 104, 132
Statistical Task Force, 48, 98, 115
Storms
 frontal, 7
 airmass, 3
Surface free energy, 17

Transformations, 43

Vonnegut, B, 17, 21, 89

Weather Modification Board, 48, 115, 139
Wilcoxon-Mann-Whitney statistic, 38, 125
Wind shear, 30